This book is a Water Training Institute *Travel Companion* and is designed to be read in its entirety during a round-trip airline flight

MLS
for **Buyers**

ALSO BY TAB EDWARDS

Paper Problems

"Coffee *is for* Closers ONLY!"

I&O: Imaging *&* Output Strategy

MPS: Managed Print Services

Lessons *of the* Navel Orange

Batman, Robin, David Beckham,
and the Naked King

Chocolate Peppers

The Art *of* Movement

.

TAB EDWARDS

MPS *for* BUYERS

MANAGED PRINT
SERVICES

Expert Insight, Recommendations,
and Best Practices for Companies
Implementing or Considering the
Purchase of MPS or other Similar
Imaging & Output Solutions

TMBE, PHILADELPHIA, PENNSYLVANIA 19129

TMBE / *The Water Group*

ISBN 978-0-9700891-8-2

This publication is designed to provide authoritative information in regards to the subject matter covered. It is sold with the understanding that the publisher is not engaged in rendering legal, accounting, or other professional services. If legal advice or other expert assistance is required, the services of a competent professional person should be sought.

—From a declaration of principles jointly adopted by a committee of the American Bar Association and a committee of publishers.

Tab Edwards books are available at special quantity discounts to use as premiums and promotions, or for use in corporate training programs. For more information, please visit the website TabEdwards.com.

Designed by Water Creative
Philadelphia, PA.

1   3   5   7   9   10   8   6   4   2

TTX

TABLE *of* CONTENTS

Introduction: Managed Print Services Turns 16 Years Old. A Look Back 15

1 / Clarifying Managed Print Services 29

2 / How Managed Print Services (Generally) Works. A Summary 88

3 / Assessment Data Gathering. Where it all Begins 104

4 / 20 Questions 129

5 / The Imaging & Output Value Spectrum 132

6 / Considerations for the Purchase and Implementation of Managed Print Services 135

Conclusion 153

About the Author 157

Index 161

MPS

for **Buyers**

INTRODUCTION

Managed Print Services Turns 16 Years Old. A Look Back

It's been approximately sixteen years since true "Managed Print Services" (MPS) as we know the solution today was introduced into the market. At its introduction, the promises were appealing: greater freedom for Information Technology (I/T) managers and their staffs; lower total cost of operation; improved reliability and uptimes; increased user satisfaction and productivity; an overall improvement in office efficiency; and, recently, greater environmental friendliness. That was the promise, but how has the solution fared from a practical standpoint? Is it delivering on its promises or is the reality of MPS something short of the mark? The reality is somewhere in the middle.

Some will argue that MPS as a solution offering is overrated and has failed to live up to the promises. I will say at the outset that the Managed Print Services solution *itself*—or at least, a true, well-run MPS program— is not solely to blame for the solution's mediocre performances at some companies; it just happens to be the innocent bystander that has been caught in the crossfire between the expectations of the MPS buyers and the program's delivery & execution by the solution providers. The majority of the fault for failed MPS implementations lies at the feet of MPS solution providers that sell "MPS programs" that are not only ill-conceived and shoddily run, but are not even true MPS programs to begin with. And buyers are not exonerated for their role in the solution's less-than-expected performance. Unrealistic expectations of the solution coupled with buyers' unpreparedness and lack of cooperation with the MPS solution provider have also contributed to the solution's so-so performance where that is in fact the case.

So, looking back, from the buyer's perspective, how has the solution actually fared over the last sixteen years? Has the solution delivered on the cost-saving promises on which it was sold? If so, what percentage of the implementations has done so? Has the solution led to the efficiency improvements that MPS sellers tout as inherent benefits of the program? Are customers "satisfied" with their decision to have purchased and implemented the solution? If so, what percentage is satisfied? Has Managed Print Services as a solution been a success? Truth is, NOBODY KNOWS! Therein lays one of the major problems with MPS: the measurement of "success" and the failure of most MPS solution providers to do so.

If you categorize Managed Print Services as an outsourcing solution—which, in the purest sense, it is—then one could make the claim that the success rate of MPS at achieving its cost-reduction objectives is the same or similar as that of outsource solutions in general. Approximately 70% of companies enter into outsourcing agreements like MPS to cut costs. Yet, only 50% of outsourcing engagements meet their financial objectives. Managed Print Services—a bundled, comprehensive solution—is a utility or *outsourced* Imaging & Output (related to print, copy, fax, scan and their associated complements and processes) arrangement that companies primarily enter into as a means to cut operational costs. And, one can argue, just like other outsourced arrangements, about half of all Managed Print Services engagements fail to deliver the financial promises on which they were sold. If so, there are several reasons. Forrester Research, a technology and market research company, reports that the top-3 causes of outsourcing failures are: poorly-defined processes, poor project management, and *no metrics for measuring success.* The latter reason is consistent with that which my colleagues and I have found to be the main reason why it is difficult to know whether MPS is delivering on its cost-savings, Return on Investment, Payback Period, or other promised financial benefits: MPS solution providers either incorrectly measure the financial and operational performance of their MPS implementations or they don't measure it at all.

Most MPS solution providers will conduct monthly, quarterly, and sometimes, annual performance reviews with their MPS clients. These reviews typically consist of a review of the number of pages produced, service level performance, and—pay close attention—a "cost-reduction review." Sounds good, right? But here is the problem: When MPS solution providers perform this type of cost-reduction review, they only measure the per-page or "click" costs that include the cost of the hardware, ink & toner, and maintenance. Then, many solution providers will compare *this* per-page cost (which some mistakenly will try to sell as the Total Cost of Ownership) to the Total Cost of Ownership costs they calculated as part of the initial office assessment. This is purely slight-of-hand. It is the same type of misrepresentation as a weight-loss guru who gains clients by touting his success at helping his clients lose weight quickly. The weight-loss guru weighs a new client on day-one with all of the client's clothes on—winter coat, boots, hat, shirt, pants, socks, and undergarments—and shows the client that he weighs 200 pounds. The guru then instructs the client to purchase a one-week supply of his miracle weight-loss drink, to drink 2 glasses of the potion each day, and to return to the guru's office after one week. When the client returns, the guru instructs the client to strip down to his boxers to be weighed. The client strips, steps on the scale, and, to his amazement, he now weighs only 190 pounds! He lost ten pounds in one week! It's a miracle! The guru then gets a testimonial from the client and uses the testimonial to "prove" that his weight-loss drinks "work" and to attract new clients.

It's easy to see in this example that the guru did not make a fair comparison between the client's day-one weight measurement process and the client's week-one measurement process; on *day*-one, the guru measured the client's weight including all of the client's winter clothing (at least ten pounds worth), and at *week*-one, the guru measured the client's weight *without* the extra ten pounds of clothing, resulting in a perceived weight-loss of ten pounds where, in actuality, the client did not lose any weight. This is what many MPS solution providers do when they report a "cost savings" to their clients during the quarterly or monthly reviews: they are comparing the "day-one" Total Cost of Ownership

(TCO) costs uncovered during the assessment process (costs that include hardware, ink & toner, support, power consumption, network IP charges, fax phone line charges, helpdesk costs, device disposal, and other TCO cost factors [TCO will be reviewed in detail later]), to the "week-one" … "TCO" … that only includes hardware, ink & toner, and support; the client's weight *without the clothing!*

To fairly, honestly, and meaningfully determine whether or not a Managed Print Services client has achieved any level of cost-reduction, financial benefit, or performance improvement, the solution provider must weigh the client with his clothes on at day-one, and weigh the client again with his clothes on at week-one. In other words, the MPS solution provider must determine the client's post-MPS implementation costs against the client's pre-MPS benchmark costs—using the same measurement methodology in both instances—to determine whether or not there has been any reduction in costs, improvement in performance, increase in user satisfaction, or a change in any performance metrics. To do this, MPS solution providers should follow a simple 5-step *Results Validation* process as outlined below.

1. Conduct the initial office assessment following a structured, systematic methodology, incorporating

the buyer-agreed Total Cost of Ownership cost factors in the calculation of TCO.

2. Create the buyer's Current Benchmark State—how things are done at the time of the assessment—including the calculation of TCO and any other financial and performance metrics.

3. After the MPS solution has been fully implemented for one year, conduct the exact same office assessment following the same methodology that was followed during the initial assessment described in Step One.

4. Create the buyer's *new* Current Benchmark State—how things are today after the solution has been in place for one year—including the calculation of TCO and any other financial and performance metrics measured in Step 2.

5. Compare the initial assessment findings with the year-one assessment findings and determine whether or not the client's costs have been reduced and if other objectives of the MPS program have been achieved.

This Results Validation approach should be a standard operating procedure for MPS solution providers. In my opinion, it is the only credible way a solution provider can determine whether or not their MPS implementation is a "success" and whether or not the client has achieved the cost-reduction benefits on which the solution was sold. Unfortunately, because only a small percentage of MPS solution providers provide this level of thoroughness, as an industry, it is very difficult to gauge whether or not Managed Print Services as a solution is delivering on its promises.

Why MPS Engagements Fail to Deliver on the Promises

Over the years, my colleagues and I have worked with companies globally on Managed Print Services initiatives, we have conducted countless MPS office assessments, and we have also conducted a significant number of Results Validation analyses for companies. Through these experiences, we have gained significant insight into the performance of MPS implementations and how successfully the solution and the solution's providers have performed against the cost-reduction forecasts that were developed during the assessment's solution development stage. Overall, we have found that approximately 68% of Managed Print Services implementations resulted in cost-reductions, while only 59% achieved the forecasted hard-dollar cost-reductions on which they were sold. This under-performance is not an indictment of the solution itself, but reflects MPS solution providers' ability to effectively execute a Managed Print Services engagement.

Through engagements wherein my colleagues and I analyze companies' MPS implementations and identify those that fail to deliver the financial and performance benefits on which they were sold and cases where the initiatives simply fail, the main reasons that we have observed for these failures are because of:

Inexperience. The buyers planning for and managing the MPS projects and the MPS solution providers with whom the buyers work have little-to-no practical experience managing and implementing complex MPS solutions and are, therefore, unaware of the pitfalls and risks inherent in the implementation; pitfalls and risks that ultimately come to fruition.

Poorly-conducted office assessments. The Current Benchmark State baseline metrics developed by the MPS solution providers to represent the buyer's company are poorly developed and flawed, rendering the data inaccurate and valueless. Because of the flawed and inaccurate assessment metrics, it becomes nearly impossible for the MPS clients to determine whether or not they have made any qualitative or quantitative improvements after a year or two of the solution's implementation.

Lack of client preparedness. One of the major buyer-influenced reasons why MPS initiatives fail is because buyers are not prepared to have the solution implemented. For instance, when a MPS contract is signed, the buyer and the solution provider will develop an implementation plan that defines how many devices will be installed at which locations on certain dates; this implementation schedule will dictate when the client will break even on its investment and how quickly the client will begin to receive the benefits—especially financial—of the MPS solution. Too often, due to a lack of communication, lack of support, lack of executive sponsorship, and poor planning, client locations scheduled to be installed with the solution are unprepared to receive the devices and, as a result, the project gets delayed, management loses faith in the project and ceases supporting it, and the solution is never fully implemented as planned. The result: the solution does not deliver on the promises on which it was sold.

Irrational client expectations (often fueled by sales representatives). As I and my colleagues have experienced over the years, lofty buyer expectations about the benefits to be received through the MPS solution is a contributor to MPS's failure to deliver on its promises. Based on the findings of an office assessment, savvy MPS sale representatives often overinflate the potential benefits revealed through the data analysis process; some will oversell the cost savings that buyers "will" receive by purchasing the sellers' MPS proposal. For example, the MPS solution provider's office assessment might reveal that the buyer could reduce its Imaging & Output (I&O) spend by 10%. However, in the interest of closing the sale and understanding that some buyers will balk at a "slim" 10% cost-reduction, the MPS sales representative will inflate the potential savings to, say, 20% in order to motivate the buyer to purchase the solution—knowing that the buyer will never receive a 20% cost savings.

In addition, because Managed Print Services had been the talk of the town for many years, some buyers erroneously believe that the solution will be a panacea for all of the buyer's company's Imaging & Output ills; it won't. The reality of MPS cost savings, for instance, is that MPS has been demonstrated (through actual Results Validation efforts) to deliver cost savings of anywhere between 2% and 45%, with a realistic expected range of between 12% and 33% depending on the level of inefficiency in the buyer's I&O office environment. If the buyer's I&O environment is in terribly-inefficient shape, the expected savings could be high; if the offices are not in bad shape, the savings could be low. So, buyers should not have any preconceived notion of what their cost savings will be until after a professional assessment has been conducted. And when you do, don't be alarmed if your company's potential cost savings are less than earth-moving.

Lack of consistent global delivery capabilities. Fact: No Managed Print Services solution provider can deliver a multi-country, global MPS implementation all by themselves; not Hewlett-Packard, not Xerox, not Ricoh, not Lexmark, not Canon, *no one*. To deliver most global MPS deals, all solution providers need human resources, so they engage partners in the local countries of implementation for delivery and management support. This is where the problems begin. When solution providers engage their local-country partners to deliver and manage MPS on their behalf in Morocco, for example, the solution provider usually has no control over the quality or consistency-of-experience delivered by that Moroccan partner. Another reality is that solution providers oftentimes provide no performance-based incentives for the in-country Moroccan partners to deliver a high quality service with the required service levels. This reliance on unmanageable partners could lead to issues with everything from solution design to pricing to unacceptable terms and conditions and poor quality.

Another issue impacting international Managed Print

Services deals that require the solution provider to engage with partners is reporting. If, for example, a U.S.-based MPS solution provider has engaged 20 different international partners to fulfill MPS delivery and support at the local-country offices of the U.S.-based solution provider's client, the U.S.-based solution providers often have no visibility into and no means of tracking device inventories, pages generated, expenditures, or even their partners' performance—not to mention the inability to consolidate all of this data into a management report. The value of this type of local-country insight can be understood when you compare MPS solution providers that operate as just described against providers like Fulton Francis Group (a global MPS solution provider specializing in international MPS service delivery and management) that use creative tools such as their iConsolidate™ customized billing & category reporting tool which tracks and provides such solution and partner performance visibility to clients in a consolidated report.

Unfavorable contracts. MPS solution providers will often structure their contract for Managed Print Services in such a way as to ensure that they make their desired profit margins over the length of the 3-, 4-, or 5-year term of the engagement; this is not unfair or unreasonable since for-profit business exist to maximize profits.

The problem arises when solution providers surreptitiously draft contracts that they know will not deliver on the expectations—financial or otherwise—of the buyer. So while the deal may appear to be financially and functionally attractive to the buyer up front, over time, it favors the vendor at the expense of the buyer.

CHAPTER 1
Clarifying Managed Print Services

Before I delve into this discussion of Managed Print Services in greater detail, it is important to provide a working definition of "Managed Print Services" as I will use the term throughout this book. The reason is simple: If your perception and understanding of Managed Print Services is different than that which I will use throughout this book, it will cause confusion as I pontificate on the solution within this text. It's sort of like a person trying to explain the intricacies of baking brownies when everyone in the audience believes the presenter is speaking about *chocolate cupcakes*: they know the discussion is about baking something sweet, but because the audience does not know exactly *what* that sweet thing is, they never receive the full benefit of what is being discussed.

What is "Managed Print Services"?

Some readers might think it unnecessary to define *Managed Print Services* considering that the solution has been on the market for sixteen years now. But, as I am reminded every time I speak to audiences about the solution, everyone has a different interpretation of what the solution is and how the solution works. And until everyone proceeds with a common definition and understanding of what makes an Imaging & Output-based solution "Managed Print Services," any in-depth discussion about the solution will in all likelihood be rendered ineffective by a lack of common understanding.

People have different definitions of what counts as a true "Managed Print Services" solution; some MPS solution providers believe the solution to be multifunctional devices/peripherals (MFPs) leased to a company on a cost-per-click basis billed monthly, while others believe the solution to be a bundle of printers plus toner leased to a company and billed on a monthly basis. Still others define the solution as toner plus maintenance bundled together and billed on a monthly basis. MPS buyers, on the other hand, might consider the solution to be something totally different than that which the MPS solution provider is selling. Such a disconnect rarely, if ever, results in a positive outcome for the MPS buyer or the seller, primarily driven by the buyer's lack of confidence in the seller's ability to deliver a smooth, working solution implementation experience ("If the sales representative doesn't even know what MPS is, then how can they help me implement it successfully?").

The best working definition for Managed Print Services—a definition that all credible, experienced, MPS solution providers and MPS clients agree with—defines *Manage Print Service*s as:

> An Imaging & Output (print, copy, fax, scan, the associated complements) offering through which the service provider takes primary responsibility for meeting the customer's office printing needs. The service provider satisfies these needs through a bundled solution that includes the hardware, supplies & consumables, service, and overall fleet management necessary to satisfy users' Imaging & Output requirements over an extended period of time.

The Benefits of MPS to the Buyer

There are many benefits MPS buyers should realize through the implementation of a professional Managed Print Services program. While, understandably, for-profit companies' primary interest in the solution is its cost-reduction benefits, there are other important benefits to be realized through the proper implementation and management of the program.

The creation of a strategy. I believe the starting point for any business initiative, including Managed Print Services, should be the development of a strategy. Diligent MPS solution providers will work with buyers to develop a MPS strategy that will guide the activities associated with the planning, implementation, management, and execution of the MPS initiative.

Hard-dollar cost savings. MPS implementations— where Results Validation efforts have been executed for validation of the solution's impact—have been shown to reduce the Total Cost of Ownership of companies' Imaging & Output environments, commonly in the range of 9% to 40%. Because the MPS landscape today is littered with "MPS" solution providers all eager to make a sale, more and more companies are offering buyers guaranteed cost savings to back-up their claim that their MPS solution *will* save the buyer money. Some of the guaranteed savings are delivered to buyers in the form of a bank check up-front (often called a "pre-bate"); some are delivered to buyers as periodic rebate payments based on the buyer achieving some predefined usage milestones; some are offered as credits against future purchases with the solution provider's company; and some are structured as fancy loans that will be recouped from the buyer's monthly MPS payments.

While the prospect of receiving guaranteed cost savings is attractive, the buyer must understand that they, too, must guarantee to perform and behave in certain ways related to their exploitation of the MPS solution. Sometimes, the requirements necessary for the buyer to receive the guaranteed cost savings are so stringent that many buyers cannot live up to the requirements and, subsequently, forego the guaranteed cost savings option.

Improved reliability & uptime. Comprehensive MPS implementations will include newer, more reliable devices in addition to preventative maintenance programs and structured service request & delivery programs, all combining to provide an Imaging & Output infrastructure that is more reliable and available to users.

Increased user satisfaction. Professional I&O assessments will often include an evaluation of user satisfaction with the Current Benchmark State of the office output environment, and the solution provider conducting the assessment will use the data as input into the solution design process in an effort to develop an offering that improves user satisfaction. Based on my study of user satisfaction data gathered during assessments, Managed Print Services has been shown to increase users' satisfaction with their MPS program by as much as 70%.

Improved user productivity. Through the streamlined, balanced device deployment models—placing the "right" devices at the "right" locations based on the usage requirements in the location—and training & transition management, users at MPS-implemented locations have been shown to perform their jobs more effectively as it relates to the use of I&O devices and user-related workflow processes.

Optimization. Every benefit derived from the implementation of Managed Print Services is a result of the solution provider's ability to develop a MPS solution model for the buyer that reduces high costs, waste, excess capacity, and inefficiency, while improving user-related workflow processes and process-driven operations.

Knowledge. Through the assessment activity, MPS buyers will (often for the first time) be made aware of the details about their office output environment: how many devices they have, which devices they have, how much the stuff costs, user frustrations, the reliability of the hardware infrastructure, operational inefficiencies, and other valuable information which the buyer did not know previously. This information helps buyers make informed business decisions about their I&O environment. An added benefit, and one that I think is highly-valuable, is that the buyer will have the opportunity to learn best-practices from the MPS solution provider as they work side-by-side toward the execution of a "successful" MPS program.

Freedom to focus on core competencies. At its core, a true Managed Print Services program is an outsource solution. As such, buyers can remove themselves from the non-valuable day-to-day activities involved in supporting and managing an Imaging & Output operation—all of which can be supported and managed by the solution provider. This frees the buyer to work on those initiatives that he or she determines to be more valuable.

Being more "Green." Through Managed Print Services, buyers are interested in reducing the cost of printing while also being environmentally responsible. Related to environmental responsibility are the four perspectives (categories) of "being green"—Preservation, Energy, Biosphere, and Global Warming—and the implications for buyers as you develop Imaging & Output strategies influenced by your company's drive toward sustainability. Managed Print Services supports these "Green" initiatives:

1. Preservation. The preservation of nature: trees, forests, plants and natural habitats from the forces of developers. MPS: Reducing paper, proper equipment disposal, and implementing a digital infrastructure to limit the need for physical pages.

2. Energy. Concern for the depletion of our energy resources. MPS: Using more energy-efficient devices, fewer devices, energy-saving mode, and implementing a digital infrastructure.

3. Biosphere. Sustainability. Whether an item's use will be depleted or sustained naturally. MPS: Recycling and working with environmentally-conscious MPS solution providers.

4. Global Warming. Greenhouse gasses that trap heat in the atmosphere and increases the earth's temperature. MPS: Using products that meet Nordic Swan and Blue Angel eco-label emission limits, and limit carbon dioxide (CO_2) emissions using printers with Instant-on Technology.

Elements of a True
Managed Print Services Solution

Given the definition of Managed Print Services provided above as a solution "to satisfy users' Imaging & Output requirements over an extended period of time," my experience in helping companies evaluate, design, implement, and manage MPS solutions revealed that the most successful, satisfying implementations all included—*at a minimum*—eight (8) solution elements. I believe these elements must be included in for the I&O solution to be considered a "true" Managed Print Services solution:

An Imaging & Output Strategy

When embarking on an Imaging & Output optimization initiative or any project that will involve an investment of time, resources, and money, the effort should be guided by the dictates of a well-reasoned strategy. Why? There are two primary reasons:

1. Every company or organization that wants to accomplish "something important" (such as the successful execution of a Managed Print Services program) needs a strategy. A strategy is a plan that defines how a company will accomplish the things ("Goals") it has determined are important to its viability. Any activities in which a company engages that are not in sup-

port of the strategy and company's goals are wasted efforts.

2. Business initiatives that are pursued under the dictates of a well-reasoned strategy are 60% more successful than those which are not.

For many companies, the word "strategy" conjures thoughts of Freddy Krueger; it's a concept so intimidating that many neglect to develop one. To them, a strategy is "something that those folks at McKinsey do for major corporations with money to burn, not for prudent companies like mine." This is a common reaction I hear when the benefits of developing a strategic plan becomes the topic of conversation. And it's understandable: many business owners have no experience with strategic planning so they don't really understand it. And since we fear things in direct proportion to our lack of understanding of them, it's no wonder that millions of business people worldwide are intimidated by the idea of developing a strategy. But rest assured: developing a strategy is not Astrodynamics. It's actually a straight-forward process of defining:

- **Goals**. A "Goal" is a broad intended outcome of an initiative or an activity in which the company is engaged.

- **Objectives**: An "Objective" is the measurable (usually quantitative) manifestation of the "Goal" which it supports. In other words, the Objective defines in measurable/quantitative (and time-specific) terms how the company will know that it has accomplished the "Goal" which the Objective supports.

- **Initiatives**: Once the Goal(s) and Objective(s) have been defined, specific "Initiatives" must be defined in support of the Objective. An Initiative is a project or other undertaking that defines what must be done in order to achieve the Objective it supports.

- **Action Plans**. Once the Initiatives have been defined, an "Action Plan" (or "Task List") must be developed which defines the specific tasks/actions that must be executed in order to complete the Initiative it supports.

Voila! Now you understand what a basic strategy is all about. This is represented in the diagram below.

STRATEGY: THE "BOGEYMAN"

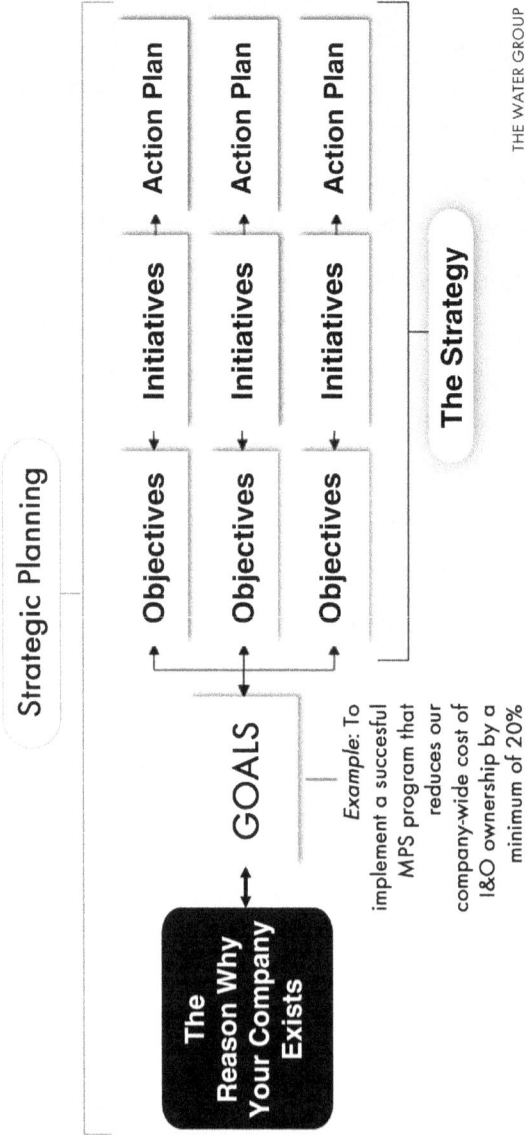

Strategic Planning

The Reason Why Your Company Exists

GOALS

Example: To implement a succesful MPS program that reduces our company-wide cost of I&O ownership by a minimum of 20%

Objectives	→	Initiatives	→	Action Plan
Objectives	→	Initiatives	→	Action Plan
Objectives	→	Initiatives	→	Action Plan

The Strategy

THE WATER GROUP

Different people have different interpretations of what a *strategy* is. A strategy can be something quite complex or something embarrassingly simple, depending on the reason why it is developed. But however you define it, if it helps you accomplish your goals and objectives in the most efficient manner, then it's a successful strategy. Consider this: A Harvard-educated mechanical engineer and a 17-year-old high-school drop-out are standing on the edge of a lake wanting to get across to the other side. The lake is approximately 100 yards across and 22-feet deep at its deepest point. There are no boats, canoes, or other floatation devices available to them; they have only the earth's natural materials available to them where they stand.

The engineer had it licked. He had developed a strategy that would surely get him across the lake. He decided that the best way to get across was to do the following: First, he would use some of the sharp-edged rocks as cutting instruments and cut the fallen tree branches into 6-foot-long pieces. He would then hollow out a tree limb, use a sharp rock to cut a hole in one of the maple trees, and insert the hollowed-out tree limb into the tree to extract sap. After that, he would rub two sticks together to create a fire (after all, every engineer was a boy scout at some point in their lives). He would use the fire to heat the sap which would then be used to bind the

6-foot-long pieces of tree branch into a leak-proof raft. Ingenious! He would then wait for the lake to calm so that his ride across it would be peaceful. The engineer estimated that—from start to finish—he would have his raft completed in approximately 4 hours. Almost bragging, the engineer asked the snotty-nosed punk drop-out how *he* planned to get across the lake. The kid replied, "I'm just gonna swim!"

The Imaging & Output Office Assessment

An a*ssessment* is a systematic approach for gathering evidence on how well a company's actual performance matches the company's expected or desired level of performance. An assessment is often referred to as a Benchmark of a company's office Imaging & Output environment—a snapshot of how the environment operates in its current state.

A robust I&O assessment effort should involve the following motions:

[1] Determine the "Study Group"

A *study group* is the collective of the buyer's general offices, departments, floors, facilities, or buildings to be assessed or studied. For large companies with many disparate buildings/office locations, it is impractical to assess the entire company's offices and other facilities that are in-scope for the assessment. The reasons are time and money: to conduct an assessment of an entire large company, the process could take a full year, by which time the data gathered at the start of the assessment process will likely be outdated and not very useful; from a cost standpoint, the *cost* (in money and time) associated with conducting such a company-wide assessment would be prohibitive (i.e. too expensive) and neither the solution provider nor the buyer would be willing to pay such steep costs. Instead, solution providers will study a subset of the buyer's company's offices and buildings, often referred to as a "representative study group."

A representative study group is a subset of the company being assessed—a subset that contains all of the usage characteristics that exist throughout the buyer's company. A usage characteristic is the general way users in a company create and produce imaging and output. If, for example, a company has marketing, accounting, finance, and I/T "usage characteristics" (the way each of these functional areas use imaging and output), the solution provider should be sure to include offices in the study group with users that have these usage characteristics in order for the study group's usage characteristics to be representative of those that exist throughout the entire company.

Many MPS buyers request (and often demand) that the solution provider study the buyer's entire company for the assessment. After all, they reason, if you study the entire company, the resulting data will be the most accurate possible. Unfortunately, that is not always the case. In addition to the reason I stated in a preceding paragraph about how year-long assessments will contain useless data that was gathered at the start of the year-long endeavor, there is a more compelling reason why a belief that "more data is always better" in an assessment is not necessarily the case, and that reason is something called *Statistical Power.*

Statistical Power explains the relationship between an increase in sample size (the size of the study group) and the resulting degree of accuracy (representativeness of the sample to the actual performance of the population/company as a whole). Basically, statistical power shows that:

- The pattern of accuracy-growth is *not* linear; the accuracy of a sample equal to half the data population size is not 50% but very near to 100%. This means that by studying only 50% of a company's offices, the degree of accuracy or representativeness of the data will be much greater than 50%.

- Good accuracy levels can be achieved at relatively

small sample sizes, provided that the samples are *representative*.

- The result of this relationship between sample size and accuracy is that, beyond a certain sample size, *the gains in accuracy are negligible, while sampling costs increase significantly*! This is illustrated in the diagram below.

Statistical Power

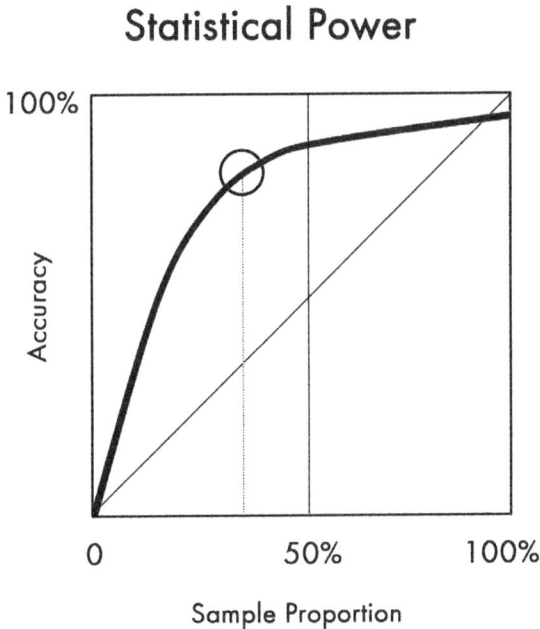

[2] Collect Cost Data to Calculate the Total Cost of Ownership (TCO)

With for-profit companies especially, and for other organizations in general, the primary benefit of implementing Managed Print Services is cost-reduction. And in order to understand and exploit the true cost reduction potential of MPS, it is important to understand the concept of Total Cost of Ownership. By definition, the Total Cost of Ownership (TCO) is the total cost of acquiring, owning, and making an Imaging & Output asset available to users over an extended period of time. TCO considers *all* of the costs associated with having and making available a company's printers, copiers, MFPs, scanners, fax machines, and their associated complements. Think about the implications: In order for a company to own a network-attached multifunctional device and to make that device available to its users, for instance, the company will incur such costs as: the device itself, toner (it can't produce printed-output without it), paper (the actual physical output from the device), power (they don't work too well if they are not plugged in), network connectivity (otherwise, the device will simply be a non-shared analog copier), phone charges (needed for sending and receiving analog fax documents), maintenance & support charges (in order to be available to users, the device must be serviced and repaired when broken), and other complementary costs necessary to make the multi-functional device available to users.

During the assessment process, the solution provider should work *with* the buyer to determine the *cost factors* that will be used to calculate the buyer's company's TCO—including *all* the costs incurred by owning and making available their Imaging & Output assets—and then define the specific costs, rates, and fees the buyer's company pays for each agreed upon cost factor (for example, the buyer's company pays $0.0670 per kilowatt hour for power consumption). There are various cost factors (e.g. paper costs and toner coverage on a page) and metrics (e.g. $0.0045 per page of paper) that can be used to calculate a company's Total Cost of Ownership for Imaging & Output. The table below provides examples of such factors. To calculate the most realistic representation of the costs a buyer's company actually incurs with its I&O assets, I recommend providing the values and calculating the costs for each of the TCO Cost Factors listed in the table. The items marked with an [*] are optional, and their inclusion in or exclusion from the company's TCO calculation is at the discretion of the buyer.

Total Cost of Ownership Cost Factors	
Business Days in an Average Month	Amortization Exclude / Expense Limit (for printers that are expensed)
Color Jobs Containing Color (must be < 100%)	Finance Costs
Toner/Ink Coverage (mono/ color)	Device Replacement Age (Years)
Toner Failure/Breakage Percentage	Hardware
Power Cost Per Kilowatt Hour	Supplies & Consumables
Work Environment	Service
Hours Per Work Day	Asset Management [*]
Days Per Work Week	Installation [*]
Weeks Per Work Year	De-installation [*]
Devices on nights and weekends? (Y/N)	Proper, Legal Disposal of Assets
% of idle time in power save mode	Asset Administration [*]
Maintenance Supplies & Labor Costs	Purchasing [*]
% Supply life (early warning)	Cost of preparing and delivering a purchase order
Paper Cost Per Page	Cost of paying an invoice
% of Pages Duplexed	Space: floor-standing devices [*]
Mfg. Date Offset to Approximate Install Date	Cost per Square Foot
Phone Line and Port Costs	IP (Network) Connectivity Cost
Avg. phone cost/ sent-fax	Print Servers
Amortization (Depreciation) Years	Helpdesk Costs
User and I/T Training [*]	

It comes as a surprise to most Purchasing, Sourcing, Facilities, Real Estate, and I/T managers when they realize that their company spends much more money on the TCO for their I&O assets than the managers realized. Most managers erroneously assume that their cost of printing, for instance, consists of the cost of the printer, the ink & toner, break-fix, and power consumption; you read it correctly: most managers even neglect to include such obvious costs as the cost of the *paper* being printed or copied, not to mention the extensive list of actually-incurred cost factors listed in the table above! For this reason, I often describe Total Cost of Ownership as "buyer enlightenment," wherein the buyers are made to understand and acknowledge that their company's total spend for its I&O assets is not only more significant than they realized, but also that the magnitude of the spend in many cases is significant enough to warrant a cost-reduction initiative like MPS in order to save the company money and help improve the company's overall profitability.

How Costs are applied to each device to determine the device's TCO and Average Cost-per-Page

Total Cost of Ownership, previously defined, is the total cost of acquiring, owing, and making an asset available to users over time. *Cost-per-page* (CPP) is calculated by dividing the TCO of a device by the number of pages produced on that device. It is a measure used primarily for comparison purposes. More and more, however, the *Average Cost-Per-Page* (ACPP) metric is being used by companies to determine the relative level of usage-efficiency in the device fleet. The Average-Cost-Per-Page is simply: (A) the average of the total population of CPP results of a particular device category. For instance, the Average Cost-Per-Page for a company's printer fleet would be the sum of all of the individual CPP results for each printer divided by the number of printers; and/or (B) the monthly TCO of a device (e.g. a printer) divided by average monthly number of pages printed on the device. The ACPP metric is very useful because, for instance, it can give you an at-a-glance view of just how (in)efficient the company's print, copy, MFP, fax, and overall hardcopy environment is operating from a cost-efficiency standpoint. If a company notices that the ACPP for its collective printer fleet is $0.10 and that the ACPP for its MFP fleet is $0.08, then the company can safely assume that the printer fleet is—on average—operating less efficiently than the MFP fleet, and that there is an opportunity for improvement with both sets of fleets, given that the ACPP in both fleets is considerably high compared to that which is considered to be generally efficient.

Some companies even use ACPP to identify and prioritize areas that require further investigation. For example, if a company uses MFPs from two different manufacturers, and, after performing an office output assessment, they notice that—all things being equal—*Manufacturer A's* device fleet has an ACPP of $0.05 while *Manufacturer B's* fleet has an ACPP of $0.10, the company will want to investigate why one manufacturer's costs are twice as high as the other's, even though both sets of devices are being used in the same manner within the same company.

When my firm conducts Imaging & Output assessments for companies, the companies' managers are often interested in understanding how the costs that we collect from their company are ultimately applied to each device to arrive at the devices' TCO and Average Cost-Per-Page values. It's rather straight forward—if you have access to a Total Cost of Ownership calculation tool. I won't burden you with an overly-detailed explanation of the process, but, simplistically, this is how it works:

1. Determine the number of months the device (such as a printer) has been installed and made available for use, and the average monthly number of pages printed on the devices.

2. The aforementioned *Total Cost of Ownership Cost Fac-*

tors are applied to the printer, usually based on the manufacturer's device specifications and the number of pages that are printed/output from the device each month; this application of the costs to the device is the purpose of a TCO tool or a spreadsheet.

3. Once the cost factors are applied to the printer, the costs are tallied, resulting in the Total Cost of Ownership for the printer.

4. The Average Cost-per-Page (ACPP) is calculated; it is simply the monthly TCO of the printer divided by average monthly number of pages printed on the device. This process is illustrated in the diagram below.

TOTAL COST PER DEVICE
and Average Cost-Per-Page

$300/yr

Total Cost Factors
Purchase Price/Depreciation
Toner & Supplies Cost per Year
Power Cost
IP Network Cost
Maintenance Cost
Phone (Fax) Cost
Helpdesk Cost
Paper Consumption Cost
Print Server Cost Allocation
Driver Installation Cost
Other

Average Monthly Pages Printed = 500
ACPP ($25/month ÷ 500pages) = $0.05

This cost-application example illustrates a very important point about a device's cost-per-page: **The cost-per-page of a device cannot be factually determined until that device is placed into service within a given company that experiences specific costs and metrics for the various cost factors that combine to determine a device's TCO and, subsequently, its Average Cost-Per-Page**.

Every day we see manufacturers' advertisements touting the idea that their brand of printer "has a lower TCO than the competitors' printers." Whenever I see such advertisements I am continually curious about them and wonder whether the manufacturers placing the Ads are simply "marketing," ignorant to the realities of TCO and CPP, or knowingly misleading consumers. Either way, it's not a flattering perception.

To fairly compare the TCO of different devices, you must first convert the TCO values into a common unit of measure; that unit of measure is (Average) Cost-Per-Page. I've always been of the opinion (an opinion supported by facts) that TCO in general, and Cost-Per-Page specifically, is something that can only truly be measured when a printer or other device is being used in normal operation. Think about it: what are the main factors that go into the calculation of a printer's TCO/CPP?

- The acquisition cost of the printer

- The cost factors and cost metrics applicable to that model of printer

- The number of pages produced on the printer

So let's assume we have two competing printers: *Printer A* and *Printer B*. And let's also assume that each printer costs the same amount of money to purchase and that they were both purchased and installed on the same date. In addition, each printer is installed in the same company (which means they have the same TCO cost factors) on the exact same table side-by-side in the mail room. For simplicity, let's say the only real difference between the two printers is the number of pages produced on each printer per month.

Printer A		Printer B	
Acquisition Cost = $2,400	$480/yr	Acquisition Cost = $2,400	$480/yr
Annual TCO Cost Factors	$600/yr	Annual TCO Cost Factors	$600/yr
Annual Pages Produced	6,000	Annual Pages Produced	4,000
Cost-Per-Page	**$0.18**	**Cost-Per-Page**	**$0.27**

Admittedly, this example is not perfect. For instance, with an actual assessment and cost calculation exercise, the number of pages produced on a device will impact the device's Preventative Maintenance Cycle (PM Cycle), potentially affecting the support and supplies costs for the device which would change the devices TCO. However, conceptually and in many cases practically, this example illustrates how it is possible for two of the exact same devices operating in the same company in the same location to have different costs per page. Taking the example further, it also illustrates how a higher-purchase-priced printer can have a lower Cost-Per-Page than a printer with a lower purchase price and consumables costs.

Manufacturers will argue that: "The TCO and CPP comparisons we advertise assume that each printer is placed in the same environment, have all of the same costs factors and metrics (except, maybe, consumables costs), and produce the same number of pages every month. They are based on side-by-side comparisons in a lab setting." To which I say: Okay, but how often does *that* happen in the real world? In the general office, people don't use printers and MFPs like that. Output production can vary greatly from machine-to-machine, and the degree to which one department maintains their devices and uses them with care can be totally offset by another department's blatant abuse of the devices. These things and other real-world anomalies will have an effect on the total cost of owning the devices and making them available to users (TCO), as well as the Average Cost-Per-Page of the pages produced on these devices.

[3] Collect Quantitative & Qualitative Data

In an Imaging & Output assessment, quantitative data are the data that are measurable and will typically be used in the calculation of costs and spend, and the determination of (in)efficiency. Examples of quantitative data include the number of users in a study group, the number of devices, the number of pages generated, the amount of time it takes a user to send a fax page or to make a copy, and various cost factors & metrics.

Qualitative data, on the other hand, are data that pertain to characteristics, conditions, and that which is not straightforwardly-quantifiable, such as user feedback, device features, infrastructure concerns, and service quality. Such qualitative data help the buyer and the solution provider identify existing problems, inefficiencies, process breakdowns, and opportunities for improvement.

I have found that any discussion of quantitative and qualitative data related to MPS and the Total Cost of Ownership ultimately leads to a discussion of "hard" costs and "soft" costs, so I will offer my thoughts on the subject.

Hard Costs vs. Soft Costs vs. Direct Costs vs. Indirect Costs

Throughout my years of working with both MPS solution providers and sellers globally, I have rarely seen an instance where the concepts of "hard" and "soft" costs were applied properly by both parties. Sure, at times, a well-schooled solution provider will correctly apply the concepts to a MPS engagement, or a savvy buyer will educate the solution provider on which is to be used when, and how each is to be used. Unfortunately, this is the exception.

As best I can recall, many years ago some consultant was hired by an early MPS solution provider to develop a systematic assessment process. And, as we consultants are sometimes inclined to do, the consultant created a detailed Total Cost of Ownership calculation and reporting method designed to impress the solution provider, and got carried away with segmenting the costs that make up TCO. The consultant segmented the cost elements into multiple strata, including "direct" costs, "indirect" costs, "hard" costs, "soft" costs, "primary" costs, "secondary" costs, and other … creative stratifications. And as it so often happens with technology, other MPS solution providers copied much of this solution provider's intellectual property, including segmenting costs into hard, soft, direct, and indirect costs. Unfortunately, as solution providers adopted this model of TCO reporting, they neglected to gain an understanding of those cost terms. If they had, they would have realized their malapropism and that such cost-nomenclature is not applicable to I&O, MPS, or hardcopy TCO. As a result, most solution providers misuse the terms today.

So, here is a quick overview of hard/soft/direct/indirect costs 101.

A "soft cost" is actually a construction industry accounting term that was used to identify costs that were not considered "direct" construction costs (hence the terms "direct" and "indirect" costs), and included the costs for such things as water delivery, interest charges, financing, and legal fees. Today, "hard" and "soft" costs are not generally used anymore in this industry; they have been replaced with the same general accounting terms everyone else uses: General & Administrative expenses.

Other industries have also had their versions of "hard" and "soft" costs. With mutual funds, for instance, "hard" dollars were things paid with cash, and "soft" dollars were things paid in-kind and through barter (e.g. you give me that sofa and I will tell my clients to use your service). And "soft costs" are even used in the solar panel industry, referring to such things as permit costs and filing paperwork to apply for a permit. Regardless from where the term "soft cost" originated or where it is commonly used today, the way it is being used with Imaging & Output TCO (and MPS specifically) is incorrect. For example, many MPS solution providers label such costs as power consumption and phone line costs (for faxing) as "indirect" or "soft" costs which is not correct. "Hard" costs are costs for which a company outlays *real* currency: cash, "foldin' money," as I refer to it. It is a cost whose payment can be traced to an actual cash outflow on a financial statement. In other words, if you have to physically transfer money to pay for something, that payment is a "hard" dollar cost. And if you *must* use the term "soft" cost—to be safe and consistent with the way most everyone else in the world considers the term—use it to describe non-monetary and time-related "costs," such as: Bob is a Human Resources Manager who earns a wage of $50 per hour. It takes Bob two minutes to walk to the copy room to copy a document and return to his desk. Therefore, the "soft" cost of Bob going to make a copy is $50 per hour x 2 minutes = $1.67. This is called a "soft" cost because Bob will not be paid an additional $1.67 in cash ("hard" dollars) each time he goes to copy a document. Such "costs" are mostly useful in building a Business Case for improving user efficiency by reducing Bob's "soft" cost of copying a document. Today, the term Document Production Costs (DPC) is used in an effort to describe the true "cost" (hard costs + soft costs) of producing and retrieving a document.

[4] Develop a Topological Map

A *topology* in the technology arena refers to the layout and networking of computers. A Topological Map is a diagram of the physical floor or office layout and relationships between the Imaging & Output assets in that office and the user who use them. It is a fancy way of describing a floor plan that identifies where each device is located on the floors of the study group in relation to the users on those floors who use the devices.

Mapping provides a visual representation of the floors (offices) being studied. It helps to visually spot problem areas and it helps the solution provider analyze user-related workflows by understanding the typical processes users in that office go through to create and retrieve output in the course of performing their jobs.

[5] Collect User Input and Feedback

One of the main reasons why Managed Print Services implementations fail is because users do not accept the solution nor do they adapt to it by using the infrastructure in the manner in which it was designed for efficiency. Therefore, it is critical to get the user community involved early in the process (even making them a part of the MPS pursuit team) and to take their input seriously.

One of the best ways to make the users feel as though they are part of the process is to solicit their input and feedback about what is working, what is not working, their frustrations with Imaging & Output, usage and usage-related workflow issues, and what they believe they need in order to help them do their jobs more efficiently (related to imaging & output). This is most easily accomplished through the use of a survey (web-based is most convenient) and a series of in-person interviews and even focus groups.

[6] Perform a User-Related Workflow Review

A user-related workflow review is an analysis of how users use the study group's I&O assets in the course of performing their jobs. It includes a *Time & Action* study which approximates the amount of *time* it takes for users to create and retrieve their printed output, to send a fax document, and to scan an image into some destination location (the *actions*).

[7] Analyze the Current Benchmark State

After the solution provider has gathered the cost, quantitative, qualitative, user, mapping, and workflow data, a snapshot of the Current Benchmark State of the buyer's company will be developed and an analysis of the benchmark state will be performed. The analysis commonly includes: a tally of the quantitative data (e.g. device count, pages printed); TCO and Average Cost-Per-Page calculations (aided by the use of a TCO tool); and device placement & floor plan inefficiencies.

From this analysis, the solution provider will create a snapshot of the Current Benchmark State of the study group environment and (assuming a representative subset of the buyer's company was assessed) extrapolate the study group findings to represent the buyer's entire company's performance. The snapshot commonly includes: the number of users in the study group and the company; the number and types of devices; the number and types of pages produced per device and throughout the environment as a whole; the TCO and ACPP per device and overall; and other metrics such as user-to-device ratios, the number of different vendors' devices and device models represented, the age of the device fleet, the percentage of network-capable devices actually attached to the network, and other such metrics.

Generally, the purpose of the assessment and the analysis of the data is to identify areas of high cost, waste, excess capacity, operational inefficiency, user frustrations, lack of device balance (e.g. too many A3 format MFPs and too few single-function network-capable printers), and environmental unfriendliness. It is only after the Current Benchmark State has been created and analyzed that a solution for improvement can be effectively developed.

[8] Design a Solution for Improvement (Recommended Future State)

After the Current Benchmark State and all of its associated metrics and findings have been developed, the solution provider will have enough information available to determine what is needed to reduce the company's costs, improve inefficiencies, reduce waste & excess capacity, and improve the overall user experience. When that determination is made, the solution provider will develop a recommended MPS solution, fleet design, and deployment model that address each of the identified areas of waste, high cost, and inefficiency; they will design a solution recommendation for improvement, or a *Recommended Future State* design.

At a minimum, the solution design developed by the MPS solution provider (with the buyer's input as appropriate) should include recommendations for:

- Consolidating devices; nearly all non-optimized companies have too many I&O devices. Also, there will always be opportunities to combine a printer, copier, and fax machine into one MFP device.

- Reducing personal (non-shared) printers; on average—as determined by the devices' ACPP—these devices are significantly more expensive based on the low volume of pages printed on them.

- Consideration of device-to-user ratios; in and of themselves, these rations mean little. However, when analyzed in relation to other indicators of inefficiency, they could have significant meaning.

- Networking all networkable devices that were purchased to be shared.

- Designing efficient user-related workflow models, resulting in less time wasted "walking paper."

- Providing the functionality requirements users need to help them perform their jobs more efficiently.

- Balancing the device deployment; ensuring that the "right" devices are located at the "right" places, in the "right" quantities.

- Replacing devices aged 5-years or older. Most leasing companies and financial institutions will only underwrite a computer asset for a term that they consider to be the asset's useful life—typically 3-5 years. And since managed imaging & output arrangements like MPS are basically embedded leases, you will typically see deal terms of 3-5 years. When I am asked by buyers about what to do with printers aged 5 years or older, my response is simple: if they work and you are happy with the output quality, reliability, and serviceability, then use them, as long as you understand that aging printers cost more to operate—in part because the supplies are more expensive and the mean-time between failure decreases with age. Workgroup printers that are more than 5 to 10 years old often have supplies costs of more than double the supplies costs of today's workgroup printers and MFPs. In addition, some of the older printers don't have the basic power management, so they are powered-on all the time, wasting electricity and leading to higher costs to operate. Older devices are also less reliable and contribute to user dissatisfaction. The result: PERSONAL PRINTERS start to mysteriously appear throughout the company!

- Reducing standalone analog fax machines. While I never recommend getting rid of all analog fax machines (primarily because users know how to use them and the machines will still work if the company's network goes down), I do recommend that solution providers look for opportunities to reduce their numbers while still providing convenient fax capability through the use of MFPs, for instance.

[9] Develop a Current Benchmark State vs. Recommended Future State Comparison

The easiest way to visualize the benefits to be received from implementing a Managed Print Services or similar Imaging & Output solution is to compare key metrics and variables between a buyer's company's Current Benchmark State (how things are done today) against the solution provider's Recommended Future State proposal (how things could be with a MPS solution in place). In my experience, the most impactful comparisons between the two states are:

- TCO and Average CPP
- Average annual TCO per user
- Average pages printed per-device-per-month
- The percentage of personal printers to the overall number of printers in the company
- Device-to-user ratios
- Average age of the device fleet or the percentage of de-

vices older than 5 years

- Number of device models and vendors; this is indicative of waste, lack of administration, and inefficiency.
- Hard dollar TCO savings and the savings percentage
- Document Production Costs (hard dollars + "productivity-based" or "soft dollars")
- Cost-Benefit Analysis, including Net Present Value, Return on Investment, Payback Period, and Internal Rate of Return. This comparison is optional.

If the comparison between the company's Current Benchmark State and the solution provider's Recommended Future State (MPS) demonstrate benefits that are not compelling, then the buyer must decide whether the solution is worth investing in (for the benefits the recommended solution *does* provide) or if there are other options for improving their company's Imaging & Output environment.

[10] Present the Findings

The final stage in the assessment process is for the MPS solution provider to present the findings of the assessment and their go-forward recommendations for improvement. This presentation typically takes the form of a written report accompanied by a PowerPoint or OpenOffice Impress style of stand-up presentation.

THE I&O ASSESSMENT CYCLE
The Water Method

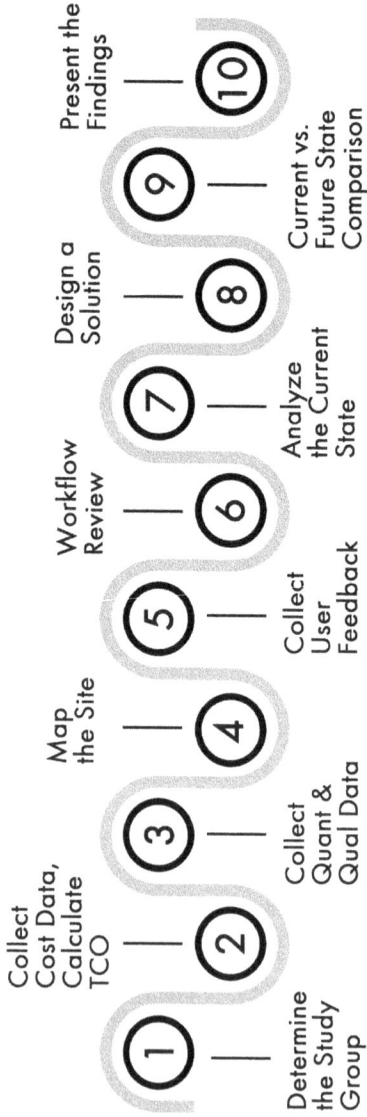

1 — Determine the Study Group

2 — Collect Cost Data, Calculate TCO

3 — Collect Quant & Qual Data

4 — Map the Site

5 — Collect User Feedback

6 — Workflow Review

7 — Analyze the Current State

8 — Design a Solution

9 — Current vs. Future State Comparison

10 — Present the Findings

It is very difficult, if at all possible, to make recommendations for improving a company's existing I&O environment without first understanding how the company is performing in its current benchmark state: does the company have excess capacity in its hardware fleet? Does the company have "too many" printers, MFPs, copiers, fax machines, and scanners? Do the company's offices operate with inefficient user-related workflow models? Is the company spending more money on I&O than they could be spending? Are the company's users satisfied with the performance and availability of the existing fleet of devices and their deployment? The answers to these types of questions can only be determined through some form of an assessment—whether rudimentary or robust—and, without gathering the information necessary to answer such questions and create the company's *Current State Snapshot*, I content that it is not possible to develop a meaningful MPS program for a company.

Why the assessment is required: An office assessment is akin to a physical examination by a physician: in order to determine what is wrong with a patient and, subsequently, what is required to make the patient better, the physician must evaluate the patient (an assessment) by *gathering evidence on how well the patient's current state of wellness matches the patient's expected or desired state of wellness.* Until the physician knows what's ailing the patient—even through the most rudimentary of evaluations (assessments)—the physician cannot make an honest, informed decision about what is needed to help the patient get well. This holds true for Imaging & Output assessments, too: until an assessment is performed of a company's I&O environment, one cannot make informed recommendations for specifically what the company must do to improve.

[*Elements of a True Managed Print Services Solution* continued]

Fleet Design

A "fleet" is the collection of hardware devices found in the current benchmark state of a company's general offices. "Fleet Design" (also called *MPS Solution Design*) is the art of analyzing a company's current state (how the company is doing things today), identifying instances of waste, inefficiency, and high costs, and creating an optimized I&O environment that reduces or eliminates the waste, improves efficiency, and has a lower cost of ownership and operation—the Total Cost of Ownership.

Total Cost of Ownership is the cost of acquiring an asset (e.g. a printer), owning the asset, and making the asset available to users over an extended period of time.

Think about the implications: companies require cash to purchase the hardware, they require cash to power it up, they require cash to provide toner & ink for the devices, they require cash to maintain the devices, they require cash to share the devices over a network, they require cash to allow fax capability, in addition to other *hard-dollar* costs (costs for which a company must write a check to pay) required to make "the assets available to users over an extended period of time."

Fleet Design involves developing a snapshot of how the company's general offices look today from an Imaging & Output standpoint, (the Current Benchmark State, including its Total Cost of Ownership), using the findings in the benchmark state (gathered from the assessment) to develop improvements to the identified problems & inefficiencies discovered, and creating a Recommended (Future) State device deployment model that optimizes the old, inefficient benchmark state of the offices. A simple illustration of an office's Current Benchmark State compared to a Recommended Future State design is provided in the illustration below.

FLEET DESIGN SAMPLE

Current Benchmark State ($100K)

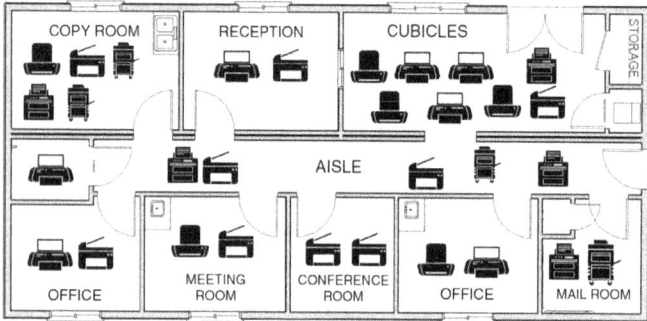

Recommended Future State ($75K)

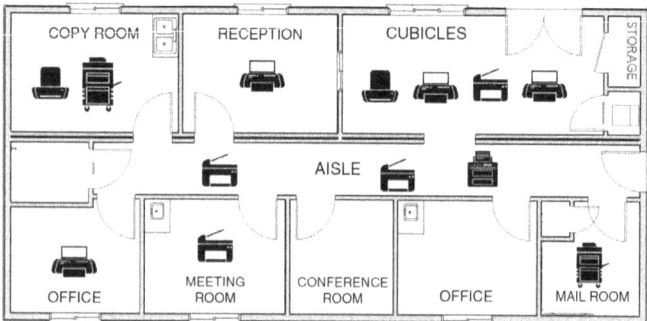

Why it is required: The Fleet Design is the optimization model for the environment that has been assessed. Without the design of an improved or optimized device fleet deployment model (including the devices recommended, the quantity of devices proposed, where the devices will physically reside in the office, how the devices are expected to be used by the users who will likely use the devices, and the total cost of the devices that will populate the Recommended Future State environment), neither the solution provider nor the buyer will know the expected cost or cost-reduction of the recommended solution, the benefits to be received by the buyer from adopting the solution provider's proposal, how the benefits will be attained, how long before the recommended solution will be implemented (and, therefore, how quickly the benefits will be received), and why the buyer should consider adopting the solution provider's recommendations.

Hardware

The hardware element of a Managed Print Services solution is obvious: a company cannot have an Imaging & Output solution like MPS without the devices that provide the imaging (on-ramp) and output (off-ramp) of content.

Supplies and Consumables

Hardware devices such as printers, MFPs, copiers, fax machines, plotters, and scanners, cannot provide their designated functions (e.g. producing printed documents or capturing digital images) with the necessary consumables like paper, toner & ink, or the necessary supplies like print heads, drum assemblies, and fusers, for example.

Service & Support

As defined above, with a true Managed Print Services offering the solution provider must take the primary responsibility for meeting the imaging and output demands of a company's users. In order to deliver this service while ensuring that the devices are available to users when needed, the solution provider must ensure that the devices are in constant operating order or that a backup option is available in the event a device becomes unavailable to users. This requires that the solution provider provides preventative maintenance, timely service, failover or backup options, and a support infrastructure (e.g. a helpdesk) to deliver a service that provides the maximum device fleet uptimes & availability possible.

In cases where companies want to take the role of self-maintainers, the MPS solution provider and the company must build a support model that integrates the company's self-maintainer support infrastructure into the overall MPS service & support model. This often involves the integration of the company's service dispatch & tracking tools into the MPS solution provider's service portal, service management application, or other tools used for tracking device break-fix cases and resolution actions. These data often become part of the solution provider's service SLAs, and for that reason, the solution providers have a vested interest in ensuring the proficiency of the buyer's self-maintenance operation.

Program Management

A MPS "Program Manager" is the MPS go-to person; it is the person at the solution provider's company who is responsible for the relationship with the key customer-MPS-contact and ensuring that the client is happy with the MPS solution. The Program Manager is responsible for the effective execution of the MPS program within a designated client company, and is also responsible for the management and administration of the overall MPS contract. If there is a problem, question, or issue with the MPS implementation or its administration, the Program Manager is the key contact for addressing such issues. Another responsibility of the Program Manager is to provide the client company with periodic (usually monthly, quarterly, and annually) updates and reporting on all aspects of the MPS solution's performance at that company, including (but not limited to) fleet utilization, efficiency improvements, changes in output production, costs, service levels, and user satisfaction. The Program Manager is also responsible for Continuous Process Improvement (CPI) related to the MPS implementation, and making recommendations to the client company that will provide cost, efficiency, and satisfaction benefits.

Why it is required: There are lots of elements of a MPS solution, ranging from solution performance to billing to service level agreements. The number of such elements makes it difficult for any buyer implementing a MPS solution to know whom the appropriate contacts are to address all of the myriad aspects of a MPS contract. For many buyers the default person to contact when all else fails is the solution provider's sales representative; after all, the sales rep owns the overall client relationship, so this is the logical choice. The problem is, however, the sales rep him or herself may not know who to contact within their own company to address an issue, resulting in delays getting a client's issues resolved. Because of the many moving parts of MPS it is logical to designate a single point of contact to manage the MPS client relationship and to serve as the contact for all things MPS. Without a Program Manager-type of role, the MPS solution provider would run the risk of creating frustrated, disgruntled customers, which could lead to the client company abandoning the MPS contract.

Fleet Management and Continuous Process Improvement

To many, "fleet management" is the act of managing an inventory of hardware devices (assets), including acquisition, tagging, installation, de-installation, MAC, logging service activities, and disposition. While I agree that all of these activities are involved in the fleet management activity, true fleet management also involves Continuous Process Improvement.

Continuous Process Improvement (CPI) is, as its name implies, the process of constantly measuring the performance of—in this case—the fleet of devices comprising the MPS program and the MPS solution deployment, identifying opportunities for efficiency improvement and cost-reduction, and making the changes necessary to exploit the opportunities.

The opportunity to continuously improve on the MPS implementation is one of the main reasons why buyers should consider adding clause in the MPS contract that allow for device movement and fleet downsizing. Fleet downsizing allows the MPS buyer to reduce the number of hardware devices (and their associated supplies and consumables) under contract in the event the client company's characteristics change for reasons such as significant employee layoffs, mergers, divestitures, or changing Imaging & Output volume needs. Say, for instance, your company signed a MPS agreement with a solution provider providing 500 printers to support your company's existing user base. During the term of the MPS agreement your company announces a layoff of 20% of its employees. In this case, it is most probable that your company would no longer need 500 printers to support its users, resulting in an environment with too many printers, excess capacity, inefficiency, and higher-than-necessary costs of operation and ownership. In such cases, buyers will benefit by including language in its MPS agreement with the solution provider allowing for *fleet downsizing*. For example:

> Under the terms of this Statement of Work, Customer may elect to change the size of the device fleet (i.e. number of devices under contract) by + or – 5% each year of the contract without penalty beginning with Year 2. An authorized representative of either Customer or Solution Provider may submit a request for change to this Statement of Work by submitting a written Change Order Request for review to an authorized representative of the other Party. Change Orders shall be the means by which Customer requests such a change to the number of devices, accessories, and hardware currently or not previously

priced and included in Attachment (X), which may be priced by Solution Provider upon request and added to the Managed Print Services set forth herein. Change Order requests shall be evaluated on an individual basis to determine any impact on pricing, payment, or other contract terms …

While such an option may bring with it added costs or other conditions by the MPS solution provider, the buyer should weigh the benefits of having such flexibility in the agreement against the potential added costs of that flexibility.

In addition to fleet downsizing, buyers should also ensure that the MPS solution gives their company the ability to capitalize on the benefits offered by any new devices and technologies that may come available during the term of the agreement, by including language that allows for *technology refresh*. For example:

From time to time, MPS Solution Provider and Customer may participate in mutually agreed upon technology review meetings to discuss new technology, software and firmware upgrades, and new hardware devices that may be appropriate for consideration as a modification or addition to the existing Fleet. Based on the results of the technology review process, MPS Solution Provider will, whenever reasonably possible, develop a proposal which includes pricing and an implementation plan, typically on a site-by-site basis for Customer's consideration. In the event that Customer would like to implement any of the new technology devices that may be available, the Parties agree to document any such agreement in a separate Statement of Work or a Change Order to an existing Statement of Work.

Why it is required: First, in order for a MPS solution provider to offer "Managed" Print Services the solution provider must actually *manage* something; that is a fundamental requirement of a true MPS offering because, without it, you simply have ... *print services*. Second—and most important—management of the device fleet and its auxiliary extensions is necessary to ensure that the buyer's MPS implementation is operating as efficiently as possible, that metrics associated with the device fleet are managed to, and that necessary adjustments which lead to ongoing improvements can be made.

Tracking and Reporting

Usage tracking can consist not only of monitoring the degree to which each printer, copier, MFP, fax machine, and scanner are being used and their associated page volumes (for page-producing devices), but can also consist of the pages being generated by characteristic (e.g. color printed pages, fax pages sent, A3 format pages produced), where the usage and pages originate, and the usage metrics needed to determine how much of the cost to charge-back to each department that uses the devices. Any such metrics the buyer considers meaningful can be included in any reporting the MPS solution provider generates related to the MPS implementation.

Tracking and reporting can be handled manually or automated, electronically and/or paper-based, software-aided, remotely, or locally. The metrics can be shared electronically using a web-based or similar portal, or paper-based using any of the means available for transmitting and sharing paper documents, including converting the documents to digital form for management and transmission.

Why it is required: If for no other reason, periodic usage tracking—whether manual or electronic—is needed in order to determine how much the solution provider should bill the buyer for the buyer's "use" (I say *use* because, with true MPS implementations, the solution provider owns the assets and lets the buyer "use" the assets for a pre-determined price) of the devices and the other elements of the MPS solution.

Single Invoice Billing

Relative to Managed Print Services, the solution provider should offer a financing option for buyers allowing them to pay for the MPS solution bundle (hardware, supplies & consumables, fleet service & support, software, tools, management, etc.) on a monthly or other term basis. One of the primary benefits to buyers of MPS is *convenience*, including the convenience of paying for all of your collective Managed Print Services through a single invoice—usually a monthly invoice. Even in cases where the buyer requests that the solution provider break out the cost of the solution by department based on the departments' specific usage of the service (i.e. departmental chargebacks), the solution provider is still, in effect, presenting the buyer with a single invoice, albeit one that is divided into subdivisions.

Depending on the pricing arrangement chosen by the buyer (e.g. Base Plus Click), the solution provider will track the buyer's MPS usage (primarily the number of pages produced) and determine how much to bill the buyer each month/period based on that usage. The most common MPS pricing models are:

Base Plus Click: The solution provider will charge the buyer a fixed monthly base charge for each device under contract, plus a variable charge based on the number of pages produced in a given month or period. This option typically has the lowest overall cost to the buyer because the fixed monthly base cost accounts for the risk of the buyer not producing the expected output volume on the installed devices; a risk incurred by the solution provider.

Pure Cost-per-Page: The solution provider will charge the buyer a monthly charge (for all MPS solution elements consumed) based solely on the number of pages produced in a given month. This pricing model is usually the most expensive because the solution provider bears all of the risk associated with the possibility that the buyer produces a number of pages each month that is below the number or pages required for the solution provider to break-even on the arrangement.

Level-Pay Billing: The solution provider will charge the buyer equal monthly payments for 12-month periods, with a resetting after each 12-month period. At the end of each 12-month period the solution provider will compare the number of pages produced by the buyer during that period against the number of pages for which the solution provider billed the buyer during that period. If the number of pages produced is *greater* than the number of pages for which the buyer was billed, the solution provider will adjust the buyer's monthly bill upward to reflect the actual number of pages the buyer produced in that 12-month period; the process will be repeated annually throughout the term of the Managed Print Services contract. The Level-Pay Billing model is the easiest to administer from a billing standpoint, and is one of the lowest cost options available due to the fact that the solution provider assumes very little risk of under-charging the buyer over the term of the contract.

Pre-Paid Pages Plus Overages: Under this model, the buyer receives a monthly allowance of pages they can produce for a predetermined fixed monthly price; any pages produced in excess of the allowance are billed on a per-page or tiered-pricing basis. This model is often seen as unattractive by buyers with fluctuating output production volumes throughout the year because, if the buyer produces a significantly higher number of pages in a given month than the contract presupposes, the buyer's costs could be much higher than budgeted for.

Why it is required: Most MPS solution providers understand that if they cannot offer buyers convenient billing options that make the acquisition of the service attractive, they will not be competitive in the crowded MPS marketplace. The reason is because—for true MPS implementations—buyers do not want to pay for the service up-front and, therefore, require a convenient, cost-effective billing arrangement to make the solution work financially and administratively within their companies.

ESSENTIAL ELEMENTS of MPS

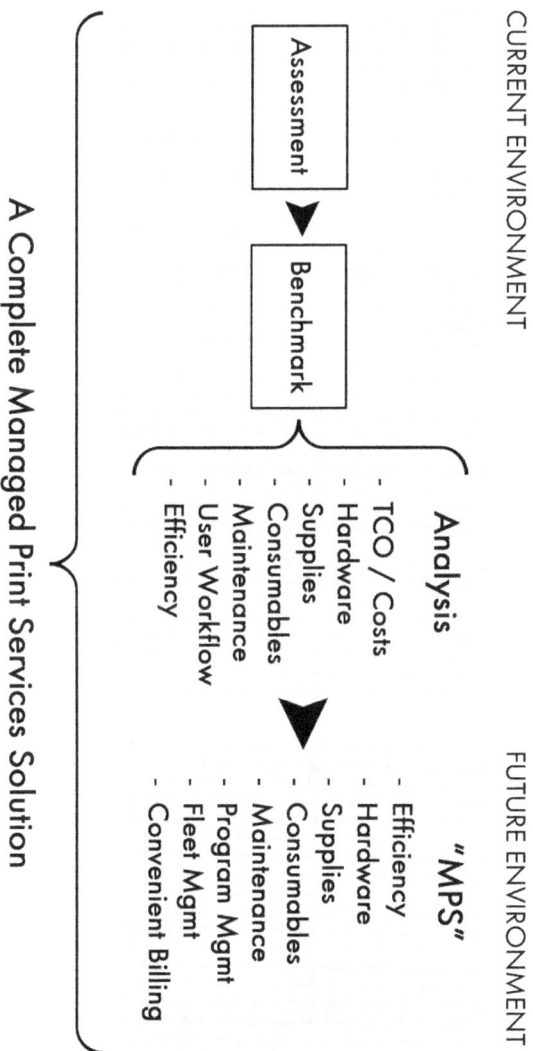

CURRENT ENVIRONMENT

FUTURE ENVIRONMENT

Assessment → Benchmark

Analysis

- TCO / Costs
- Hardware
- Supplies
- Consumables
- Maintenance
- User Workflow
- Efficiency

"MPS"

- Efficiency
- Hardware
- Supplies
- Consumables
- Maintenance
- Program Mgmt
- Fleet Mgmt
- Convenient Billing

A Complete Managed Print Services Solution

Other MPS Service Options

In addition to the service elements presented above which should be required for a Managed Print Services solution to be considered genuine, there are also (often optional) services that a solution provider can incorporate into their version of a MPS solution bundle to add additional value for the buyer, including:

- *Existing Fleet Administration*: Managing and administering the buyer's existing fleet of "out of scope" Imaging & Output devices outside of (or, in many case, incorporated into) the MPS solution.

- *Asset Management*: Providing and updating an accurate inventory of all the hardware and software assets in the buyer's company, including nodes without an IP address.

- *Phone Support*: Providing supplemental phone support—related to the MPS implementation—to offload the call burden of the buyer's helpdesk staff or to provide "Level-0" support.

- *Software*: Providing software applications that augment the buyer's MPS experience, including software applications that track remote, non-network-attached users' print output or applications that re-route users' print jobs to a lower-cost or working device.

- *Installation / De-Installation*: Including the removal of old devices, disposing of them "properly," and/or getting monetary value for them on the open market or within the solution provider's company directly. The service also includes the staging, un-boxing, driver installation, network connectivity, and physical installation of new devices.

- *Deployment Project Management*: Providing Project Management services to help ensure a well-managed MPS implementation process, from assessment through installation.

- *Training / Transition Management*: One of the major reasons why MPS implementations fail is because they are not adopted by the users. Transition Management is the process of developing and executing a plan to move the buyer's company and users from the current environment to the MPS environment; this always includes user and I/T training.

- *Ongoing Optimization*: Providing management services to ensure that the MPS implementation is continually operating at its optimal level of efficiency.

CHAPTER 2

How Managed Print Services (Generally) Works. A Summary in Ten Stages

The details of what Managed Print Services is and the various elements of a MPS solution were provided in some level of detail in the preceding paragraphs. To bring it all together, I will provide a simple outline summarizing how the solution usually works.

Different MPS solution providers not only have different interpretations of what *their* version of Managed Print Services entails, but they also have different approaches for how they execute their version of MPS. Fundamentally, an effectively-executed MPS implementation will follow a ten-stage process as outlined below. While this is a high-level overview of how the solution generally works (in light of the fact that MPS solution providers have different flavors of MPS), I have found that it is enough of an overview to give the reader a good understanding of the major facets of a functional MPS offering.

1. **A strategy** defining the goals and objectives for the initiative.

2. **The assessment** to gain intelligence about the buyer's Imaging & Output environment. At a minimum, the MPS buyer should expect the following from the assessment effort:

   - *A Findings & Recommendations Report.* This report, whether delivered in traditional report format or presentation-slide format, should provide a comprehensive review of everything that occurred, was discovered, or was determined during the assessment process, including: The assessment methodology, general findings (quantitative and qualitative), the Current Benchmark State definition and analysis, recommendations for improvement, user productivity and qualitative analysis (including user feedback and satisfaction levels), the Recommended Future State (including the recommended hardware), a comparison between the Current Benchmark State and the Recommended Future State (quantitative and qualitative), TCO analysis, cost savings and financial analyses, the Business Case (depending on the solution provider's level of expertise), and the steps to get from the current state to the desired state. **The buyer should expect to pay** the solution provider for conducting such an assessment, especially if the buyer wants to use or share the Findings & Recommendations report in any manner they wish, including sharing the data with other companies.

- *Collaboration.* The solution provider and the buyer should work together toward the successful accomplishment of all assessment-related activities, including stages of the solution development process.

- *Development of the full list of TCO factors to be used.* In order for the Total Cost of Ownership data to be as reflective of reality as possible, the buyer must work with the solution provider to determine not only the cost factors that will be used to calculate the buyer's total costs (e.g. toner costs, paper costs), but also the specific cost variables (e.g. paper cost = $0.0045 per page) to be used for each cost factor.

- *Building the Business Case.* Depending on the level of business acumen and experience of the solution provider, the buyer should expect to work with the solution provider to develop a Business Case for the MPS project. A Business Case is a management proposition for determining which investment to make among the many projects under consideration. Building the Business Case ensures that: the Purpose of the project is at the forefront of the initiative; potential risks have been considered and panned for; the planned technology is viable; the cost-benefit analysis projects positive returns; the implementation timeline has been approximated and is reasonable; and a results-validation plan has been developed. These are just a few suggestions of things to be included in the process.

3. **Development of the Current Benchmark State**: Once data have been gathered and the assessment has been completed, the solution provider will calculate the Total Cost of Ownership and the opportunities for improvement

4. **An optimization effort** to define the improvements that would address the problems identified in the benchmark state.

5. **Development of the Recommended Future State**, showing what an improved Imaging & Output environment would look and perform like.

6. **Implementation of the MPS solution.** After all contracts and agreements have been signed and project & transition management begun, the implementation process begins with what is referred to as *Discovery & Design* (D&D), whereby the solution provider analyzes—on-site—each office that is targeted to receive the Managed Print Services solution. During the analysis (which is like a mini-assessment), the solution provider will identify the existing devices and certain metrics for each, such as the device model, device features, the number of pages produced on each device, and the physical location of the devices (the *discovery*), and will then *design* a fleet deployment model that optimizes the office's I&O infrastructure. After the buyer signs-off on the solution provider's recommended fleet design, the devices will be shipped; the old devices de-installed and disposed of, the new devices installed, and the MPS program implemented.

7. **Program Management**, to ensure that the buyer's MPS implementation is operating as efficiently as possible.

8. **Tracking & Reporting**, to gather metrics that can be used for billing, reporting, and decision making.

9. **Monthly billing** or some other convenient, periodic billing interval.

10. **Fleet Management and Continuous Process & Improvement**. After all, the solution is called *Managed* Print Services.

HOW MANAGED PRINT SERVICES WORKS

START

Strategy

Assessment

Current Benchmark State

Optimization Effort

Recommended Future State

Go, No-Go

Implementation

Program Management

Tracking & Reporting

Monthly Billing

Management & CPI

From Where the MPS Cost Savings Emanate

Most everyone acknowledges that Managed Print Services—when executed properly—saves companies money (hard dollars) by reducing the cost of ownership. What many people do not know, however, is where the cost savings come from. Though the distribution of cost reductions can vary from solution provider to solution provider based on how they manage their MPS programs, most will agree that the primary drivers of cost reduction with MPS are device & output consolidation, improved device reliability and uptimes, and waste reduction through more efficient processes.

Device and output consolidation

There are three forms of device consolidation that directly impact the cost of Imaging & Output ownership: hardware consolidation (combining multiple devices of different types into a single, multifunctional device), fleet consolidation (combining multiple devices of the same type into fewer devices of the same type), and page output consolidation (directing print pages from inefficient print devices to more efficient, cost-effective devices).

Most companies that have not been optimized have "too much" *stuff*: too many printers, too many MFPs, too many copiers, too many analog fax machines, too many print servers, too many unusable toner cartridges, too much wasted paper, too much … *stuff*. By consolidating all of this stuff through an optimization effort (which should always precede a MPS implementation), companies can realize significant cost savings. Say, for instance, a company has two identical network-attached printers in close proximity supporting the same user pool for general office printing. Also let's assume that each of these printers should "optimally" produce 5,000 pages per month, for a total of 10,000 pages combined. If the Total Cost of Ownership of each printer is $200 per month, and each printer only produces 1,000 pages per month, then the Average Cost-Per-Page of each printer would be $0.20; very expensive. So, by consolidating the two printers and routing users' print jobs to the one consolidated printer, the buyer's TCO for both printers combined would decrease from $400 per month to $200 per month, and the ACPP would decrease from $0.20 to $0.10 per page; not great, but much less expensive and more efficient.

DEVICE CONSOLIDATION
and TCO Reduction

$200/mo $200/mo

1,000 pp/mo   1,000 pp/mo
ACPP = $0.20   ACPP = $0.20

Consolidated TCO = $200/mo; ACPP = $0.10

The same logic applies to combining multiple disparate devices into one common device. For example, a company with two analog fax machines sitting on a table next to a networked workhorse printer and an old analog copier—each producing a paltry number of pages per month—can reduce its waste, excess capacity, and cost of ownership by combining a fax machine, the printer, and the copier into one MFP.

Improved device reliability and uptimes

Generally, newer Imaging & Output devices are more reliable than "old" devices. The mean-time between failure for devices 5-years or older, for instance, decreases each year the devices are in use. They become less reliable with frequent downtime, often requiring frustration-inducing maintenance and service calls in order to bring the devices back to a usable state.

Unusable or broken devices not only increase maintenance costs, but also paper waste. How? When user send a print job to a printer and that printer is not functioning, the users will re-print the document to the same printer first (as one study has shown to be the case), and when they notice that the second print job did not print, the users will send the print job to a different printer. Then, when the broken printer becomes functional again, the users' print jobs that were sitting in the queue will then print, creating wasted pages because the users no longer need them, other users using the now-revitalized printer will discard the other user's documents, and/or many of the printed documents will simply get lost.

THE PAPER TRAIL
What Ultimately Happens to Paper Produced in the Office?

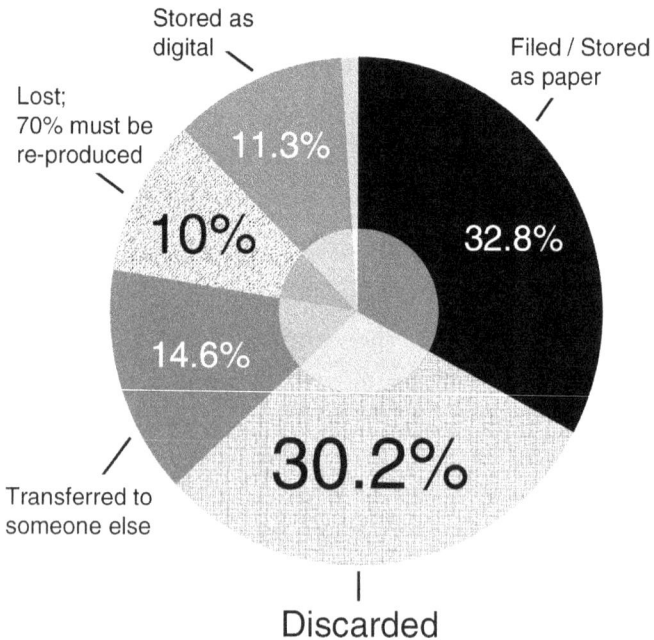

Stored as digital

Filed / Stored as paper

Lost; 70% must be re-produced

11.3%

32.8%

10%

14.6%

30.2%

Transferred to someone else

Discarded

SOURCE: Marketing & Business Integration (MBI), LLC.
Study: *Follow That Paper!*, 2005

Another cost-reduction and user satisfaction benefit of improved device reliability and uptimes is the reduction in the number of helpdesk support calls and their associated costs. There are studies that show that between 25% and 33% of all helpdesk calls are printer-related calls. For buyers who do not know the number of printer-related helpdesk calls their company receives annually or the amount of money spent on helpdesk calls, there are a few different ways to approximate that number, including: (1) approximating the number of printer-related helpdesk calls made for each printer annually, and (2) approximating the average number of printer-related helpdesk calls each user make per year.

At some companies, especially larger unmanaged corporations, I have seen where, on average, 5 calls are made for each printer in the study group. At other companies, I have seen where each user in the study group places an average of 4 printer-related helpdesk calls per year. So, in the former case, if the company has 3,000 total printers, that company's helpdesk likely received 15,000 printer-related helpdesk calls in a year (3,000 printers x 5 calls/printer/year). In the latter case, assuming the company has 10,000 users, the company likely received 40,000 calls to its helpdesk that were printer-related. In this case, if we assume an average cost of a helpdesk call to be $25, then the company could be spending $1,000,000 on printer-related helpdesk calls. [The average cost of a helpdesk call varies by company. I have worked with companies whose company-provided average cost of a helpdesk call was as low as $9 per call, and I have worked with others where the company-provided cost of a helpdesk call was $35].

Through the pre-MPS assessment process, solution providers will measure device reliability and service levels as best they can. They will use the information to design a solution that provides more reliable devices; improved preventative maintenance and service delivery processes which will improve device uptimes and availability; and increased buyer-acceptable service levels.

Waste-reduction through more efficient processes

Whether it is streamlined billing processes, ordering processes, service request processes, job-routing processes, user training processes, device-driver installation processes, or device recycling process, improved processes can not only improve operational efficiency, but also reduce the hard-dollar cost of ownership. Take, for example, toner ordering, fulfillment, delivery, and recycling.

I would be willing to bet that if yours is a company of more than 1,000 office employees, you will have unusable or unnecessary still-in-the-box toner cartridges sitting in a copy room, mail room, or in closet someplace;

most companies of a certain size do. This often happens when companies' users have the autonomy to purchase "office supplies" without executive approval as long as the cost of the item(s) is less than $1,000. Nowadays, you can get a pretty hefty network-capable printer for less than that—not to mention inexpensive ink-based printers—and users know that fact. So, it is inevitable that printers and their associated ink & toner (including "backup" quantities purchased) will start showing up at these companies. Over time, there will be so many untracked printers—from every hardware manufacturer—throughout the company, and so many spare toner & ink cartridges for these untracked printers, it becomes inevitable that toner and ink will continue to be ordered for devices for which "backup" toner & ink have already been ordered. Over time, as printers break, are discarded, or are replaced with different printers that do not use the same toner and ink as that which has been stockpiled in the supplies closet, the company will be left with toner and ink cartridges for devices which the company no longer has. This type of waste can cost a company hundreds of thousands of dollars.

MPS solution providers with solid, robust Managed Print Services programs will implement solutions that streamline various processes which can help companies improve reduce waste and reduce the cost of ownership.

One such process is automated toner fulfillment, whereby the MPS client can elect to place an order for ink & toner directly through the MPS solution provider's ordering portal (or the client's Intranet linked to the solution provider's ordering system) or the client can work with the solution provider to develop an automated replenishment program. With an automated toner replenishment & fulfillment program, the solution provider will track the client's print usage for each device and, when a device's toner levels reach some threshold such as 10% or 20% remaining (as determined by the number of pages printed against the toner cartridge's page yield), the solution provider's system will automatically trigger an order for a replacement toner cartridge for the device. And, depending on the services offered by the MPS solution provider and the buyer's willingness to pay for it, when the new toner order arrives at the client's loading dock, it can be taken to a designated holding location (and in some cases, desk-side) where it will be installed by a representative of the solution provider. Such replenishment & fulfillment services will reduce wasted toner, significantly reducing the client's hard-dollar toner costs.

Cost-Reduction Savings Distribution

When exploring the primary drivers of hard-dollar cost-reduction with Managed Print Services programs—device & output consolidation, improved device reliability and uptimes, and waste reduction through more efficient processes—analysis of The Water Group's historical Results Validation data reveals the following distribution of the specific areas of cost savings that emanate from each of the cost-reduction drivers.

COST-REDUCTION DISTRIBUTION
From Where MPS Cost-Reductions Emanate

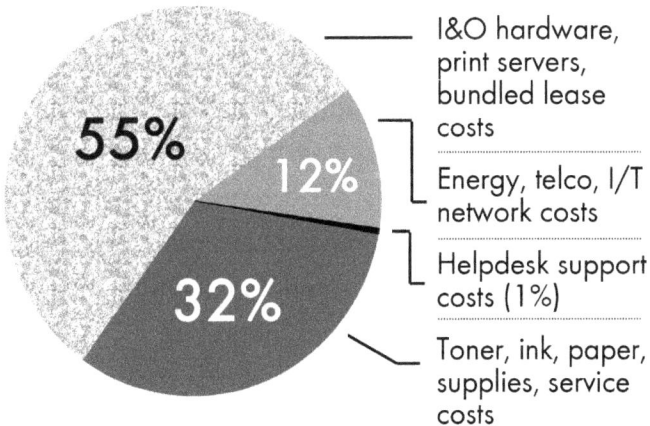

55%

12%

32%

I&O hardware, print servers, bundled lease costs

Energy, telco, I/T network costs

Helpdesk support costs (1%)

Toner, ink, paper, supplies, service costs

CHAPTER 3

Assessment Data Gathering. Where it all Begins

In my experience, the most important aspect of the Managed Print Services "process" is the Imaging & Output assessment. The reason is because *everything* that happens during a MPS campaign is dictated by the assessment: the potential cost savings are determined; the efficiency improvements are determined; the number and types of devices required are identified; the project plan is dictated; the transition and training needs are dictated; the roles & responsibilities are dictated; the implementation plan is dictated; the payback period on the buyer's investment is dictated; *everything* is dictated by the assessment, including the success or failure of the MPS project! That is why the quality of the assessment that is performed is extremely important. And when it comes to the assessment, the data gathering activity can make-or-break an assessment. The reason is because the key assessment data-gathering metrics—device inventory and page-count volumes—are those upon which all else is ultimately based. When solution providers or buyers conduct an assessment, there are two primary approaches for gathering these metrics: automated and manual.

The *automated* approach to usage-tracking (page-count volume and characteristics) and data-gathering typically involve the use of a data collection agent, such as a software application, which is installed in the buyer's environment and tracks and captures the number and types of pages being produced in the study group (e.g. mono, color, A3, A4, duplex, etc.) during the course of the data gathering activity—usually 30 days.

The *manual* approach to determining device usage and page-counts is an approach I developed back in 1999—out of necessity—which has since become the de facto standard approach that solution providers have adopted and still use to perform the data gathering activity in assessments. The reason I created the process was $220,000. Back in 1999 I was engaged by a major global corporation to conduct an Imaging & Output assessment of their entire U.S. operation—all 50,000 users across multiple geographic locations—as a prelude to a Managed Print Services project. At the time, my company used a data collection agent (software) that we installed onto our clients' computer servers in order to track print usage for 30 to 60 days. This client, however, was not having it; they did not allow 3rd party software applications to be installed within their corporate firewall. They offered me an ultimatum: either conduct the assessment (for which my company was to be paid $220,000) without using the unallowable software agent or get lost. That day, I went back to my office and figured out a way to conduct an Imaging & Output assessment (which I called the "Hardcopy Operational Assessment") without the use of software; it would be conducted manually. What I discovered was that the manual approach to data-gathering could not only be performed more quickly than 30 to 60 days, but also yielded results that were more reflective of the client's actual page volume performance. The rest, as they say, is history.

The manual data-gathering process is a process for, among other things, determining the average monthly number of pages printed/output per device in the study group. To do this, the provider identifies the device's installation date (date of first use) and the total number of pages printed on the device since it has been installed;

rather straight-forward. So, for example, if a device has been installed for ten months and over that ten-month period it has produced 10,000 pages, then the average number of monthly pages printed on that device would be estimated at 1,000 (10,000 pages divided by 10 months).

But how accurate are the results that each method yields? After all, the degree of accuracy of each method's extracted data will determine the degree of accuracy of other aspects of the MPS project, including the buyer's estimated Total Cost of Ownership and potential cost savings. The answer to the "accuracy" question is: something less than 100%.

As with anything mathematical, there is room for statistical error in each data-gathering approach. For instance, with the automated approach to data-gathering, collecting data during a non-representative month (months of the year where printed output and device usage are not typical) will skew the page volume totals used to represent the annual total number of pages produced throughout the study group. And with the manual approach, if a printer has been installed for 10 years, for example, then the manually-calculated monthly page volume average could be slightly off if we assume that printer usage and the associated pages decrease each year, which it has for the last several years.

Two Primary Approaches to Data-Gathering: An Analysis

Automated Data Collection

I used two calculations for determining the degree of accuracy of the print volume data gathered using automated data collection agents over a 30-day or 60-day period. The first was calculating the probability that the *month* selected for the 30-day collection period is the month whose print output production is representative of the actual average of all 12 months' print output production. The second was calculating the degree of accuracy of the resulting print volume *data* at representing what the actual average monthly output production is over a full "usage cycle." A *usage cycle* is a full calendar year's (12 months) worth of device usage data and the associated number of pages generated from those devices over that period.

There are 12 months in a calendar year. For simplicity, let's assume that the average month is 30 days-long with 22 working days. There is a direct correlation between the number of people working in an office and the number of pages that are printed in that office; if there are 100 people working in an office, there will be more pages printed in that office than if there were 50 people working in the office. In addition, the average number of pages consumed by workers can vary depending on the type of organization in which they work. I categorize companies along a paper-consumption spectrum starting with *Conservative Paper Consumers* on the low-volume end, and *Prosperous Paper Consumers* at the high end of the spectrum. Conservative Paper Consumers (CPCs) are dominated by smaller companies & organizations, and compete in such in such vertical industries as Life Sciences, retail, and manufacturing, while Prosperous Paper Consumers (PPCs) compete in high-paper-intensive industries like financial services, legal, Marketing, and consulting. According to the American Paper Institute, each U.S. worker uses about 5,400 sheets of paper a year. Other research, including that conducted by The Water Group, shows that the typical PPC consumes an average of approximately 14,100 pages annually. Assessment engagements that I and my colleagues have personally conducted over the years show the average number of pages consumed per-user annually across all industry segments along the paper-consumption spectrum is about 8,695 pages per year (at an annual cost of approximately $740 per user in TCO), and at the low end of the spectrum, the average number of pages consumed annually per user cost approximately $459 per user per year.

PAPER CONSUMPTION SPECTRUM
Average Annual Print Consumption Per User (Pages)

Low

Average

High

(5,400)

8,695

14,100

$740
per user, per year

Another interesting finding from our research is that the number of pages printed or consumed in a typical office—regardless of the industry—varies from month, to month. This is to be expected if for no other reason than the correlation between the number of people in an office and the number of printed pages consumed. And since there are fewer workers in a typical office during certain months than there are in others, there will be fewer pages consumed in these certain months than in others. Think about it logically: December (holidays), January (holidays and New Year), February (short month), June, July, August (summer vacations), November (Thanksgiving). These events mean that there will be fewer workers in the office during these low-worker months, which translates into fewer people producing printed and copied output, which means the output produced in these months will typically be less than the output produced in a standard month where users are in the office for each of the 22 business days of the month. In addition, companies in certain industries produce significantly higher page output volumes in certain months than in a "typical" month due to the nature of the business. For example, financial services firms will increase their print output at the ends of each calendar quarter, and accounting firms will experience print output increases during income tax season.

These data show that, contrary to what some people may believe, companies with traditional office environments rarely, if ever, print the same—or even approximately the same—number of pages each month; I have reviewed sixteen years' worth of actual assessment data that support this finding.

So, what is the probability that by sampling paper output production for any one month (30 days) of the year, you will get a result that is equal to or representative of the actual average monthly output production of a full year (usage cycle)? The answer: technically, zero (because no companies' users produce the *exact* same number of pages every month) and at best, 17% if we assume there are two months in a year where the print output volume is an average of each of the twelve months' print output volume. Looking at the empirical data, there is actually one month (30-day period) of the year that is most representative of a true monthly average print output volume for companies in general, and that is March-April.

The problem with collecting sample usage data for 30- or 60-day periods using an automated data collection agent (or any other method, for that matter) is that solution providers don't wait until that one time per year (March-April) when the print output volume is most representative of the buyer's true average monthly consumption; most solution providers erroneously assume that users produce the same amount of print output each month. So I pose the question thusly: What is the probability of selecting one month (30 days) out of twelve in which to collect usage data and selecting that one month that represents the buyer's true average monthly print output production? Using the formula for *Probability* (*p*) below, we can calculate the likelihood.

In this problem, there is a set of "N" elements (12 months) and a sub-set of "n" *favorable* elements (the one "correct" month during which print output production is a truly an average of all 12 months), where "n" is less than or equal to "N."

$$p = \frac{n}{N}$$

That means the probability of selecting the one month of the year where the users' print output production is an average of all twelve months' is 1 out of 12 (8.3%). So, if there is only an 8.3% chance that the month you track print usage for 30 days is the "correct" month, that means there is a 92% chance that you will track an un-representative or "incorrect" month.

The second part of the calculation is determining the degree of accuracy of the resulting 30-days-worth of output production data in reflecting the true monthly average print output production (i.e. the *sample mean*). The only way to perform this calculation is to know (or to be able to approximate) the *actual* average number of print or hardcopy (print, copy, and fax) pages produced in a typical general office for every month of the year. This part can be difficult without any data to use as your frame of reference. To find an answer, I have compiled data from the myriad office output assessments my colleagues and I have performed over the years, and I have calculated the average number of pages printed per-month-per-device—for each month of the calendar year—across *all* of the companies we assessed. These data support conventional wisdom that output production varies by month in the typical general office environment. I have also calculated the number of months where that specific month's print output production is within 5% (the margin of error that still leaves the data within a statistically-significant confidence level of 95%) of the actual monthly average (the *mean*). The result: Only one month's actual print output production was within 5% of the true mean (monthly average) of all twelve months' print output production. In other words, the probability that any random 30 days-worth of data collected using the automated data collection software agent approach or any other method fell within 5% of the actual monthly value was also only 8.3%.

The formula below can be used to determine the probability that both the "correct" month was selected to collect usage data *and* that a random 30-day sample collected is within 5% of the actual monthly value.

$$P \left(X \otimes Y \right)$$

In the equation, "X," the probability of selecting the correct month to study (assuming there is one month out of 12 where that month's print output volume is representative of the true year's average in that study group) = 1/12 or 8.3%; and "Y," the probability that the data you ultimately come back with is within 5% of the actual mean = 1/12 or 8.3%.

Applying the formula to determine the probability: P (X (x) Y) = (0.083) * (0.083) = 0.7%

This calculation shows that the probability of getting a monthly print output production value that is statistically significant (representative of the actual monthly output production volume in the environment with 95% confidence) through randomly sampling 30 days-worth of data using automated data collection (or other) method is only 0.7%, or less than a 1% chance.

Following is an example. The diagram below will help illustrate the key points being made through the above scenario.

EXAMPLE: SAMPLING ERROR
When the Monthly Volume is Based on a 30-day Random Sample

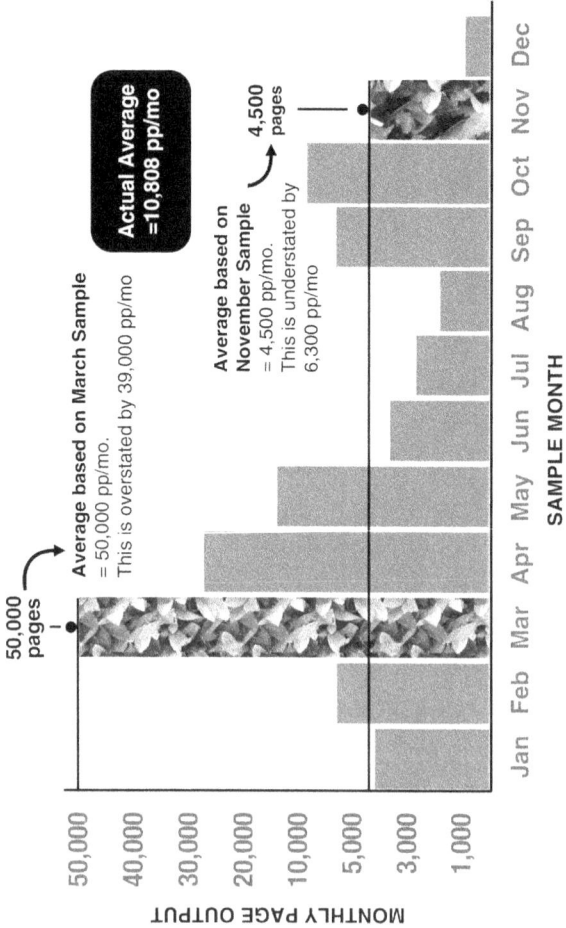

Actual Average
=10,808 pp/mo

50,000 pages

Average based on March Sample
= 50,000 pp/mo.
This is overstated by 39,000 pp/mo

Average based on November Sample
= 4,500 pp/mo.
This is understated by 6,300 pp/mo

4,500 pages

MONTHLY PAGE OUTPUT

50,000
40,000
30,000
20,000
10,000
5,000
3,000
1,000

Jan Feb Mar Apr May Jun Jul Aug Sep Oct Nov Dec

SAMPLE MONTH

Assume, based on the timing of the MPS initiative for which an assessment is being conducted, the solution provider just so happens to track print output for the month of March and extrapolates that data to represent the full year. In this scenario (illustrated by the diagram above), the solution provider would conclude that the buyer's average monthly print volume is 50,000 pages and the annual volume is 600,000 pages. This is obviously significantly overstated from the actual average print volume of 10,808 pages per month as shown in the diagram. The margin of error with this 30-day sampling approach will most always be high because the same sampling error would exist regardless of which month the solution provider randomly tracks—unless, of course, the solution provider happens to track the exact month that *is* reflective of the true average (i.e. gets lucky). But the solution provider has no way of knowing which month that is—if, indeed, there is a month that is truly representative of the average within the buyer's study group.

This same logic holds true whether the data are collected for 30-day periods using automated tools or manually. Some solution providers are becoming aware of the fact that whether they collect data for random 30-, 60-, or 90-day periods, the same sampling error will occur. As a result, you will see some solution providers use the automated data collection agents to capture data for only *2 weeks*. By doing so, they argue, the data won't be much less accurate than if they collected data for 30 days, and they will be able to wrap-up the data-gathering activity more quickly, resulting in a faster turnaround process for the assessment's Findings & Recommendations report.

The problem with this automated data-gathering approach is that the tracking period of 30 or even 60 days is not long enough to yield results that are representative of the study group's true average usage. Tracking software was intended to be used over time to accomplish the following Imaging & Output-related tasks: job accounting, fleet balancing, tracking usage volume for controlling use & abuse, determining least-cost routing, and identifying chargebacks. In that capacity, it is quite useful. However, it was not intended to be used for 30-day print volume sampling to be used in assessment studies.

Manual Data Collection

As with the automated method to data collection, there are also two factors necessary for determining the degree of accuracy of the data gathered using the manual approach:

• The probability that the on-site device installation date

is approximately 2.2 months (a metric I defined which will be explained in the paragraphs below) from the manufacture date of the device as determined by the manufacture date stamped on the device or included in the device's configuration.

- The assumption that the *rate* of paper output produced in the general office decreases by the estimated average of 6%. Although paper consumption increases each year, it is increasing at a *decreasing rate* due to such factors as a decrease in the average age of the workforce, a decrease in workforce sizes, and a move to digital solutions. Due to these factors, most companies believe their overall printed and copied output *rates* go down slightly each year and, correctly, that their per-user consumption increases slightly each year.

Network-capable, shared printers (departmental, workgroup, team-shared, etc.) and convenience copiers/MFPs are not typically manufactured and stored in warehouses or on retail stores' shelves for long periods. In this era of technology manufacturing-on-demand and just-in-time inventory replenishment, manufacturers try to reduce the amount of time devices are stored in inventory in order to reduce inventory carrying costs. Manufacturers want to increase their Inventory Turnover Rate or "inventory turns," a common measure used to determine the cost-effectiveness of a company's inventory which is calculated by dividing Cost of Goods Sold by Average Inventory—both of which are common financial statement values.

A while back, I took a look at three of the top printer/MFP manufacturers in the U.S. and calculated the average inventory turn of the three companies combined. My analysis showed that the average inventory turn across all three manufacturers was 6.7; the fewest turns by one of the manufacturers were 4, which meant that equipment was moved out of that manufacturer's warehouse every 3 months on average. That result meant that, on average, each company moved print output devices (including MFPs) from their inventory about 7 times per-year for all devices, meaning every 51 days (less than 2 months). I stated earlier that my manual data gathering method assumed an average of about 2.2 months from the time a print device is manufactured until it arrives at an end-user company's location. The calculation of Average Inventory Turnover supports this assumption based on actual manufacturer data I also gathered from one of the manufacturers. In addition, my own data from actual company assessments I performed where I compared *actual* printer installation dates (when they were available from the clients' asset management reports) to my 2.2-month average estimate shows that my 2.2-month estimate is a very good predictor of the actual installation date: the actual install-data data showed an average installation date that was approximately 2.4 months from the devices' identified manufacture dates.

Therefore, the probability that an estimate of the actual installation month can be determined using manufacturing dates (in months) plus 2.2 months can be calculated as follows: The *actual* range of months-in-inventory for a network-capable printing device is between 1.7 and 3 months; the average being 2.3 months. So, the probability that using an average installation date of 2.2 months from the manufacture date is within 5% of the actual mean (average) of 2.3 is close to 100%.

For completeness, we must also calculate the following: What is the probability that choosing *manufacture date plus 2.2 months* as the estimated installation date is statistically significant, given the actual (typical) manufacturers' ship-to-loading dock measures of within 1 month, 2 months, or 3 months? The answer would be a one-in-three chance, or P=1/3, or 33%. So, you can summarize by saying that the 2.2-months-from-the-manufacture-date estimate is between 33% and approaching 100% accurate. In my opinion, however, the reality is closer to 33% than anything approaching 100%.

The second part of the accuracy determination for the manual approach is the estimated rate-of-decrease in paper output production of 6% annually. My study data reveal that 42% of the printers in companies, on average, are 5-years old or older. Since 5 years (moving to 3 years) is the standard depreciable life of a printer asset, I will use this 5-year depreciation value for the calculation. My study data also reveal that the average number of pages-per-printer-per-month in companies, on average, is 3,489 pages. If the rate of paper output production in the general office decreases by 6% annually, then the average monthly pages-per-printer values would be as follows:

Year1: 3,920 pages
Year 2: 3,698 pages
Year 3: 3,489 pages
Year 4: 3,280 pages
Year 5: 3,083 pages

So, in this example, the probability that the *average* pages-per-device-per-month of 3,489 is within 5% of the actual mean (average) when taking the decrease in office print output growth into consideration is 100%. In other words, the 6% rate of decrease in pages produced will not significantly affect the average monthly pages-per-device using the manual data collection method because this method already calculates the average using the correct formula for determining the true average of multiple variables:

Average = (X1 + Xn/N) where "X" is the variable (the average monthly page values each year) and "N" is the number of observations (5 years in this example).

Yes, the reality of the page-volume data-collection activity of the Imaging & Output assessment is that it's not perfect, but it's the best we've got. There are, however, things the solution provider, the buyer, or any other entity conducting an assessment can do to increase the degree of accuracy of the page-volume data gathered during an assessment. With the automated data-gathering approach, for instance—whereby print usage is tracked for 30 or 60 days—the solution provider can use historical assessment data across all industries and for companies of all sizes to determine the average print output consumption per-user for each calendar month of the year. Using these averages to represent the typical (or the true average) monthly print output volumes by month and by user per month, the solution provider can then adjust the page volumes they collect up or down—depending on the month(s) during which they track print usage—to more closely reflect the "typical" average month's usage based on historical data. And if the solution provider has enough of an historical repository of assessment data to determine an average by industry and/or company size, the data can be adjusted to be even more reliable.

With the manual method, the solution provider can (1) use any actual asset-tracking data they gain access to while conducting assessments (which shows the exact install dates of the devices in the study group environment), and use this collective data to adjust their estimated install dates to become more reflective of that company's actuals; and (2) gain insight from the buyer on how the user population and/or print volumes have changed over the past 5, 10, and 15 years, and use this insight to adjust their collected print output volumes accordingly.

By performing such simple-yet-effective adjustment activities, the reliability of the print output volumes gathered using either data-gathering method can be greatly improved.

Data Gathering Methods: The Pros and Cons of Each

Now that I have reviewed the two primary methods— automated and manual—used for approximating the monthly print output volume for an assessment, an obvious question is: which method is better? As you can imagine, the answer is: it depends. Many factors come into play which will give rise to one approach being more favorable to the other, such as the skill level of the solution provider conducting the assessment; the amount of time the buyer is willing to allow between the start of the assessment process and the delivery of the Findings & Recommendations report; the comfort level the buyer has with the representativeness of the data developed from each method; the buyer's preference for one approach over the other; the time-cost relationship of each method; the resources the buyer is willing and/or able to commit to the effort; and other factors.

While I believe there is merit to using both approaches simultaneously—for example, using the manual method for its speed and accuracy, and supplementing the activity with data gathered using automated data collection agents—many buyers believe there to be redundancy to doing so, and will elect to use one process over the other. In order to provide some insight into the efficacy of each data-gathering method, I will offer my thoughts on the relative merits of each method in the table below. This comparison might be useful to you as you consider which method to employ for your Imaging & Output assessment.

Assessment Data Gathering	
Automated Process: Pros	**Automated Process: Cons**
Broader reach. The use of data a collection agent (DCA) will allow the solution provider to capture page counts and other data from a large user base without the need to have people physically visit each location to do so. **Rich page output data.** Many DCAs can capture various characteristics of the pages produced in a study group, such as paper size, mono or color, the user's name, IP address of the printing device, and the number of pages printed per job. **It's technology-based.** Because the method is technology-based, some buyers might believe its results to somehow be "better" than those based on a manual approach; this makes them comfortable with the process. **Easy integration into a TCO tool.** Since the data are captured electronically, they can be fed directly into a TCO tool which reduces the potential for transcription errors, and the TCO tool will use the data to calculate various metrics such as costs. **Vetting for the MPS program**. Many solution providers use their DCA as a solution element in their MPS program. For example, it can be used for tracking print usage for billing purposes. So installing the software during the assessment process will give the buyer an opportunity to vet the software before it goes live as part of the MPS program.	**Security and virus concerns**. Many companies have become more cautious about allowing 3rd party software into their environment due to the potential or viruses and security breach. **Representativeness of the data gathered**. Randomly sampling page volume data for 30- or 60-day terms could yield results that are not the most representative of the typical average usage through the study group. **Leads to a longer assessment duration**. A typical turnaround time for an Imaging & Output assessment is 30 days plus the duration of the page volume collection activity. If the page tracking takes 60 days, it would typically take 3 months to complete the assessment and deliver the Findings & Recommendations. **Resource commitment for IT testing.** Companies will not allow 3rd party software to be introduced into their environment without vetting the software for safety and compatibility. This requires an investment of I/T resources. **Redundancy**. Most solution providers that use DCAs to track data automatically still must visit the study group locations and manually touch each device (including gathering usage data) as they would using the manual method.

Assessment Data Gathering	
Manual Process: Pros	**Manual Process: Cons**
More representative data. The print output volume data is more representative of a true average month's usage than that based on using a DCA to collect usage for random 30-day periods. **Faster completion time**. Depending on the size of the study group and the number of locations, the page count collection activity can be completed in as little as one day. This would enable the solution provider to complete the entire assessment and deliver the Findings & Recommendations report up to 30 days sooner than an automated approach that tracks page counts for 30 days. **Easy.** The method is easy to follow and teach others, which makes it easy to include more people in the data-gathering effort. **No I/T system access required**. The method does not require software to be evaluated, scheduled for testing, or installed. This eases some I/T managers' fear of viruses being introduced and intrusion.	**Potential for transcription error**. After manually collecting page volume data and (as some do) writing it down on paper data collection sheets, the data must be re-keyed into a TCO tool presenting the potential for transcription errors, unless the data are electronically entered into a spreadsheet or other tool during the data-gathering process. **Reach**. Since the manual data process requires people-resources to complete the print output volume data-gathering effort, it would require more people and expense to gather data from far-reaching locations than it would by using a DCA.

CHAPTER 4
20 Questions

One of the clearest indications of an un-managed, inefficient, higher-than-necessary cost Imaging & Output environment is one about which the I/T, strategic sourcing, purchasing, procurement, and real estate managers do not have a firm grasp on some of the basic characteristics of their organizations' important metrics and measures. For example, if a buyer does not know the company's annual spend in *total costs* (as in TCO) for all of the I&O hardware and its complements, then there is a good chance the environment is not as efficient as it could be made to be.

I developed a series of twenty (20) questions that, based on the buyer's knowledge of the answers, will give the buyer insight into whether or not there is a need to consider an initiative to optimize the environment and reap the benefits that such an optimization effort could bring. The process is simple: for each question, honestly … *honestly* … answer "yes" or "no" depending on your knowledge of your company's Imaging & Output environment. If, at the end, you are unsatisfied with your answers (which will be an indication of the condition and at-risk nature of your environment) then you might want to consider investigating the potential benefits you could receive from conducting an assessment that will provide the information being asked for in the 20 Questions which, in turn, can help you make more informed decisions about making improvements.

1. Does the company know how much money (hard dollars) they spend annually in total costs for print, copy, fax, support, use, and management of the Imaging & Output environment (not simply the cost for printers, MFPs, and toner/ink)?

2. Does the company know—within 5%—the number of printers, MFPs, copiers, and fax machines throughout the company?

3. Does the company know the current printer-to-user ratio (the number of printers per user)?

4. Do the individual departments have budgets (e.g. up to $1,000) and the autonomy to purchase technology (e.g. printers) as long as the cost is within their allowed budget amount?

5. Do more than 25% of the users have their own personal (non-shared) printers?

6. Are there more than four (4) different manufacturers' MFPs represented/installed in the company?

7. Are less than 90% of the company's MFPs attached to the network?

8. Do more than 50% of the company's MFPs provide 11" x 17" (A3) on-glass format?

9. Does the company have toner products on-hand for devices that the company no longer uses?

10. Does the company have printer, MFP, or copier leases longer than 4-year terms?

11. Is more than 33% of the company's printer fleet older than 5 years?

12. Are more than 30% of the company's help desk calls printer-related?

13. Do fax machines comprise more than 20% of the company's total Imaging & Output devices (print-

ers, copiers, fax machines, scanners)?

14. Have the company's users been formally trained on the effective use of the company's MFPs?

15. Has there been a compelling event (e.g. a flood, fire, earthquake, merger, acquisition, move, new location, etc.) mandating the acquisition of new or additional I&O equipment?

16. Has the company issued (or are they planning to issue) a RFP/RFI for hardware or MPS-type services?

17. Does the company acknowledge the need for optimization or improvement of its I&O environment, but there is no money available for an improvement initiative?

18. Does the company have MFPs, copiers, or printers with leases expiring within 12 months?

19. Is the company undergoing a cost-cutting initiative that impacts all departments?

20. Is or has the company engaged in an Imaging & Output assessment?

CHAPTER 5
The Imaging & Output Value Spectrum

The Imaging & Output Value Spectrum (the "Value Spectrum") is a graphical conception I developed to help buyers and sellers more easily grasp the relationship between the various Imaging & Output transactional product and solution options, the value and business impact of these options to buyers, the value and business impact to the solution provider, and the alignment between the interests of the buyer and solution provider. This alignment is important because it illustrates the degree to which the solution provider is forced to be more consultative in their approach to working with the buyer. Being consultative ensures the strongest collaborative working relationship. It means *not* asking "What can I sell to this buyer?" but instead, asking "What is the business objective the buyer is trying to accomplish by considering this project and how can I help them accomplish it?"

The Value Spectrum illustrates the range of possible I&O-based products, complements, solutions, and their associated value to the buyer and the solution provider. It shows the alignment between the interests of the buyer and solution provider as well as the relationship between solution-comprehensiveness and business impact. For example, a stand-alone printer (which sits at the low end of the Value Spectrum) has very little business value to a buyer and very little business impact on the solution provider. But as you move down the Value Spectrum from a single-function printer (or other I&O hardware device) toward a full Managed Print Services solution, the relative value and business impact increase for both the buyer and solution provider. This common interest will almost force the buyer and solution provider to work collaboratively toward a success project to ensure that both receive the maximum value from the solution; this is alignment.

Naturally, there are probably one hundred possible products, services, and solutions that could lie along the Value Spectrum. However, in order to illustrate the concept of the Value Spectrum, I have limited the number of I&O solution elements that lay along the spectrum as illustrated in the diagram below.

IMAGING & OUTPUT VALUE SPECTRUM

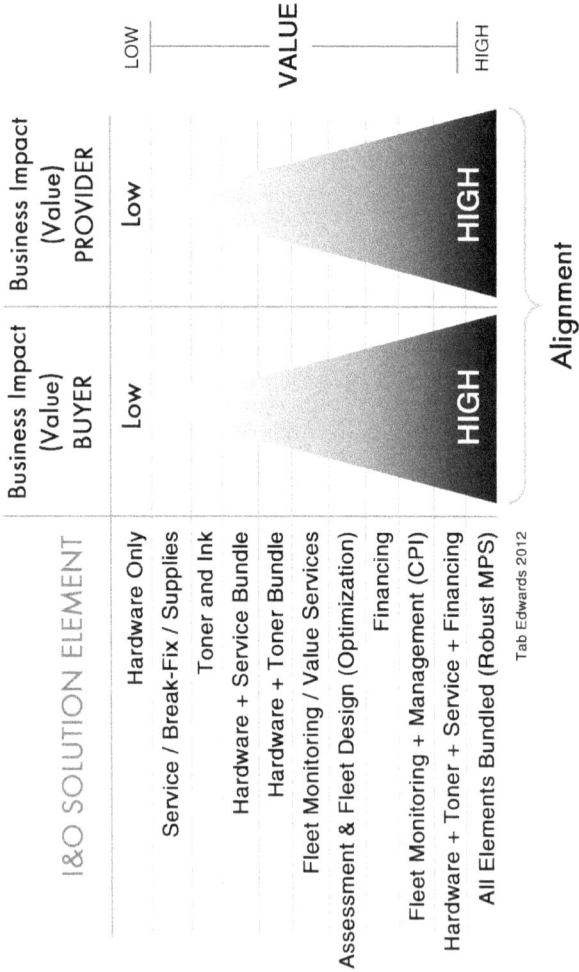

I&O SOLUTION ELEMENT	Business Impact (Value) BUYER	Business Impact (Value) PROVIDER
Hardware Only	Low	Low
Service / Break-Fix / Supplies		
Toner and Ink		
Hardware + Service Bundle		
Hardware + Toner Bundle		
Fleet Monitoring / Value Services		
Assessment & Fleet Design (Optimization)		
Financing		
Fleet Monitoring + Management (CPI)		
Hardware + Toner + Service + Financing		
All Elements Bundled (Robust MPS)	HIGH	HIGH

VALUE — LOW / HIGH

Alignment

Tab Edwards 2012

CHAPTER 6

Considerations for the Purchase & Implementation of Managed Print Services

General Considerations

Companies can improve the "success" rate of Managed Print Services implementations—with "success" being defined as the achievement of the stated objectives of the initiative—through the following best-practices and considerations:

- Development of a Strategy. A strategy will ensure that the project is being implemented to meet certain important goals and accomplish specific objectives. It will also provide the buyer with a report card on how effective the MPS solution is at delivering on the objectives for which it was undertaken. A strategy drives alignment of focus and activity among all parties to the MPS effort, and serves as the road map for how the solution will deliver value to the buyer's company.

- Gaining a comprehensive understanding of the current state base-lining and MPS assessment methodologies. This will keep the solution provider honest and ensure that the buyer establishes solid baseline metrics that can be used to determine whether or not the buyer's company has made improvements through the MPS implementation.

- Building contingencies into the MPS agreement that hold the MPS solution provider accountable for failing to deliver on the promised benefits (e.g. specific cost savings, service levels).

- Considering objectives that go beyond the quantitative (such as operating cost savings) to include operational efficiencies and the buyer's and users' acquisition of process knowledge, relevant skills, and/or expertise from the solution provider.

- Building a Business Case to ensure that the MPS project has the greatest chance of not only gaining the support of the buyer's executive management team, but also the greatest chance of succeeding. The Business Case development process forces the buyer and solution provider to consider and address every important aspect of a project such as MPS; aspects that contribute most heavily to a project's success or failure.

- Understanding the type of MPS financing deal (lease) you are entering into. The type of financing arrangement buyers enter into can have an impact on a company's expenses, accounting treatment of the hardware assets, and even earnings. The most common MPS lease financing arrangements include:

  - *Operating Lease.* With an operating lease—under the accountancy rules applied in many countries—the leased asset (e.g. printers, MFPs) generally does not appear on the buyer's balance sheet. With an operating lease, the buyer is truly paying for the use of the equipment, not its ownership. This is important, because many MPS buyers continue to act as though they continue to own the assets. The reality under this lease arrangement is that the buyer only *uses* the assets.

  - *One-Dollar Buyout.* This option—which allows buyers to purchase the leased hardware for $1 at the end of the lease term—is most suitable for buyers that want balance sheet financing that helps the buyer's company improve earnings (specifically, Earnings Before Interest, Taxes, Depreciation, and Amortization—EBITDA). This lease structure allows the buyer to capitalize all of their printing devices.

- *Master Lease / Finance Lease.* Finance leases are especially advantageous for buyers who plan to keep the equipment at the end of the lease term. In a finance lease, the assets generally appear on the buyer's balance sheet and are amortized over the life of the lease.

- Evaluating the proficiency of the solution provider. Today, it seems as though every company whose business does anything remotely associated with a printer or toner offers some form of a MPS solution. Some are good, and some are not. Sure, every one of these solution providers can look good when they respond to shoddily-written boilerplate RFPs and RFIs, but to quickly and easily get a handle on whether or not the solution provider that you, the buyer, are considering is worth a second visit, simply ask them the following fundamental questions. Every MPS solution provider that is experienced should be able to effectively and convincingly (based on your standards) answer these questions. If they cannot, then you should consider additional evaluation criteria or …

— How well do you understand Managed Print Services? Can you explain everything that happens throughout the process from how the assessment works, to how TCO is calculated, to how you determine the fleet design for locations that were not part of the assessment study group, to what happens when new printers arrive to replace the old ones, to how you will bill individual departments if needed?

— What is your company's process for determining whether or not we achieved the TCO reductions on which the solution will be sold? How will you compare the pre-MPS environment to the one-year-old MPS environment?

— How does your company handle global implementations in countries where your company does not have local offices? Also ask about how the solution provider engages local in-country partners and how they measure their international partners' performance, coordinate activities, provide common reporting, etc.

— How will you calculate our Total Cost of Ownership, Return on Investment, and the number of months before my company breaks-even on our investment in MPS?

— Describe your company's process for Building the Business Case, which elements are included, and what my role as the buyer will be in the process.

The Multi-Year Cost Savings Illusion

Although I have covered this topic in greater detail in the book *MPS: Managed Print Services*, it is such an important and necessary issue to understand for both the buyer and the solution provider—but especially the buyer—that I will share my thoughts on the matter here as well.

As a prospective MPS buyer, you have probably been presented with a multi-year MPS proposal that shows how you will save large sums of money *every year* of the agreement by moving to a MPS solution. It happens all the time. The seller's pitch goes something like this: "Dear Buyer, by signing this 3-year MPS agreement with my company, we will save you $1M per year for each of the 3 years of the agreement, for a total of $3M dollars, resulting in a 50% annual cost savings from your current spend of $2M!" And this is how the solution provider likely calculated these $1M annual cost savings: They conducted an assessment which found that the company could reduce its operating costs (from a TCO perspective) by $1M by implementing a MPS solution. Then, the solution provider presumes that, since this is a 3-year deal, the company will save that $1M *each year of the agreement* for a total of $3M dollars. Wrong!

It doesn't work that way. This example is outlined below.

- The solution provider has proposed a 3-year MPS deal

- An assessment revealed that the company spends $2M per year on Imaging & Output TCO in their current un-managed, inefficient, non-MPS environment.

- The solution provider's proposed MPS solution will cost the buyer $1M, yielding a savings of 50% *for each year of the MPS agreement.*

- The entire MPS solution will be implemented within Year 1.

This is faulty logic because, once the buyer implements the MPS solution in Year 1, **the buyer will no longer be spending $2M per year on Imaging & Output in Years 2 and 3! They will only be spending the $1M cost of the MPS solution!** This logic implies that two mutually-exclusive activities—the buyer spending $2M for MPS in Years 2 and 3, and spending $1M for MPS in years 2 and 3—will occur once the buyer purchases the MPS solution. They won't, nor can they.

What the sales rep is *really* saying is that the *Expected Opportunity Loss* (EOL) of not moving to a MPS solution is $3M over three years, *not* the *actual* expected cost savings. EOL is a concept used to express the amount of money (potentially) lost by not selecting the best solution between two options: the status-quo and Managed Print Services. That's a lot different than saying the buyer *will save* $1M per year for 3 years, which assumes the buyer will continue to save $1M each year, which is not the case in this example or in similar real-world scenarios. The reason is once the buyer's environment has been optimized with MPS implemented, the buyer's TCO should no longer be $2M annually.

A more plausible proposition from the solution provider would be to say that the buyer is spending $2M today for Imaging & Output, and tomorrow, after the buyer has implemented MPS, the buyer will only be spending $1M, for a total savings of 50%; the total dollar savings will be $1M, and possibly more through Continuous Process Improvement efforts. This logic is more realistic since Years 2 and 3 will become the norm and will not be comparable to any "better" solution proposal, which means there will not be further "significant" cost savings, save for the small future cost savings that will result from CPI.

When a buyer is presented with such a promise, I advise the buyer to ask the solution provider to re-calculate the proposal's 3-year cost savings by creating a deployment/installation schedule which shows what percentage of the MPS solution will be installed across the targeted locations by when, along with the TCO at each stage of the installation process. This should correctly reflect the fact that as more of the MPS solution is installed, the less money the buyer will have to pay for Imaging & Output, to the point where the MPS solution is fully implemented.

Buyer Bad Habits

Managed Print Services solution providers are not the only parties complicit in a poor MPS experience. MPS buyers often play a role, too. Having worked on both sides of the table—I have placed lots of MPS solutions with buyers, and I have helps lots of buyers evaluate and make decisions about acquiring MPS solutions from vendors—I have seen, for years, the little things buyers do that solution providers hate; things that impact the effectiveness of the solution provider's MPS program at the buyers' companies. These buyer bad habits can turn into MPS pitfalls if the impact of the actions is magnified.

- Fragmented Imaging & Output responsibility ("Fragmentation"). In most corporations, multiple departments have authority and profit & loss responsibility that overlap in the office Imaging & Output environment. For instance, where there lacks a single department responsible for optimal deployment, the I/T, facilities & operations, sourcing, purchasing, and user departments may each have the ability to purchase printers, MFPs, and supplies—while doing so under no overall deployment strategy. Fragmentation leads to internal conflict between the I/T and facilities/operations departments. This conflict and dis-coordination often restricts buyers from providing the necessary coordinated effort (a joint effort involving all departments that have I&O responsibilities) for ensuring a successful MPS effort, which results in less than optimal MPS implementations. In the end, the solution provider catches the heat for overseeing a failed MPS engagement.

- Unwillingness to fully support the data-gathering effort. Throughout a MPS engagement, especially at the beginning assessment stage, the solution will need the buyer to gather information that only the buyer has access to, such as the specific cost variables to be applied to the TCO cost factors when calculating the buyer's TCO. As a result, the buyer and seller either must rely on industry-standard data (which is fine) or use no data at all (which is not fine).

- The inability to provide people resources. During the MPS process, the solution provider will develop a Responsibility Matrix which identifies the people resources needed from the buyer's company and from the solution provider in order to affect a smooth process. Oftentimes, buyers will not or cannot commit the necessary resources to complete certain tasks, which either delays the project or forces the solution provider to provide the needed resources; an action that could increase the cost of the solution to the buyer.

- Failure to involve the user community up front. As I described earlier, one of the main reasons why MPS solutions fail to live up to their potential is because users do not embrace or exploit the program. My research shows that the key determinant of whether or not the user community embraces MPS (as opposed to considering it to be something that was forced onto them or an initiative designed to "take our printers away") is whether or not they were involved in the process and/ or their voices and concerns were heard.

- Lackluster internal communication about the project. A buyer's failure to do something as simple as sending out an email message or other internal communication announcing and explaining the assessment and/ or the MPS project can create suspicion and concerns among the users in the study group—concerns that always seem to find their way to the CXO community in the company. The lack of effective communication at each stage of the project has been shown to put users on the defensive against those "printer thieves," which can prohibit the solution provider from doing the best job possible toward a successful assessment and MPS effort.

- Insistence on the use of a specific vendor's hardware. I have seen and worked with many buyers who care an awful lot about which vendor's hardware is used to comprise the MPS hardware fleet. Sure, sometimes it makes sense for the solution provider to use Vendor A's hardware since 90% of the buyer's existing hardware fleet consists of Vendor A's products (it relates to user familiarity, supportability, the buyer's company's relationship with Vendor A's company, etc.). When it comes to MPS, buyers should focus less on which hardware is being used (after all, you are just *using* it, you won't own it) and more on such things as service-level agreements, uptime, reliability, usability, quality of output, operating costs, and other factors that dictate the success or failure of a Managed Print Services initiative. This is the same challenge with buyers that issue Managed Print Services RFPs that focus heavily on specific manufacturers' devices and not the major benefits to be received by MPS or the determinants of a successful MPS engagement, such as profit improvement, cost savings, ROI, efficiency improvements, user productivity gains, and user satisfaction.

- False perceptions. Partly due to the over-eagerness of the solution providers to sell their MPS solution, many buyers will allow solution providers to visit their companies to conduct (free) Imaging & Output assessments as a prelude to pursuing a MPS engagement. Buyers will do this knowing that the buyer's company has not approved such a project, the project has no funding approved, no executive support, and a snowball's chance of ever materializing into a real company initiative. Knowing this, some buyers will still allow the solution provider to conduct the assessment, and will even marshal some internal company resources to help (waste time) with the effort. These instances are unfortunate for the buyer and the solution provider: the buyer has wasted co-workers' time and other resources to support the null effort, and the solution provider wasted their peoples' time and other resource performing the activity for which there is no opportunity at the end.

Imaging & Output Security

As crippling virus and security breaches threaten the stability of companies globally, it is more important than ever for buyers to consider the security implications of entering into a MPS program; a program that often involves the use of the solution providers' 3rd party software to be integrated into the buyers' internal infrastructure.

There are various security considerations for Imaging & Output solutions along the Value Spectrum, especially MPS. This includes:

Print Security

When it comes to Imaging & Output security, buyers must consider both the legal requirements that your company must adhere to related to paper output and/ or digitization of documents (for example, certain documents should only be modifiable by certain parties and under certain circumstances), as well as the potential for confidential/classified documents getting into the wrong hands. The primary types of security threats buyers must consider include:

- *Print output*: Confidential documents that are left lying on the printer output tray for anyone to see. This is the main argument that users make for why they have and need personal, non-shared printers.

- *Network sniffing*: Unauthorized, non-company hackers "sniffing" the network in hopes of printing buyers' documents at their location.

- *Device theft*: Someone stealing either the printer or the printer hard disk to obtain valuable information that may be stored on the device.

- *Hacking and re-routing*: Hackers can access MFPs, LAN fax machines, and Digital Senders, and configure them by getting around weak security controls, to send buyers' internal documents to their e-mail addresses, fax machines, or printers.

- *Network penetration*: Software can be installed in printers with modems in them, thus allowing the hackers access to other information on that same internal network.

Considerations for securing the Imaging & Output environment

- *Secure the product.* Install devices that have the ability to prevent unauthorized users from accessing confidential information in the printer or its output tray, or from changing the product's configuration.

- *Protect information on the network.* Make sure the devices support the latest encryption and communication security protocols, such as IPsec.

- *Monitor and track the device fleet*: This will allow the buyer to identify any anomalies and/or suspicious activity that can then be further investigated.

- *Document security*: For buyers that are concerned about counterfeiting and sensitive documents like contracts, checks, etc., there are third parties that offer document security solutions that include security fonts, anti-copy paper, and security-toner that "bleeds" when the original document is altered.

- *Access Control.* Access Control printing solutions requires authentication at the printer before a print job is released. One such example is security or *private* print where users must input a 4-digit pin code at the printer before it releases the document. There is also software that buyers can use to track who printed what and where print jobs go. These solutions also allow buyers to restrict the type of machines users can print to, and they can limit the number of pages printed by user if needed.

CONCLUSION

Managed Print Services is a complex solution offering that requires the combined efforts of engaged buyers and knowledgeable solution providers to get "right." When a MPS implementation works well, it is a tremendous solution delivering business-improving benefits to buyers. But when it fails, it can be dramatic, costing both buyers and sellers many-thousands of dollars, lost time that cannot be recovered, and—at its worst—lost jobs. That is why it is very important for buyers that are investigating and considering the implementation of Managed Print Services to understand what it is, how it works, why it succeeds, why it fails, what to expect, best-practices, common pitfalls, and the consequences of failure. This is the reason why I wrote this book: to serve as a guide for helping buyers navigate the complexities of pursuing Managed Print Services.

Over the years, I have performed practically every function involved in the MPS pursuit and support efforts, including (but not limited to): the creation of MPS programs; creation of an industry-adopted Imaging & Output assessment methodology; MPS project management, program management, and deliver; MPS assessment execution, financial analysis, solution design, and implementation; and serving as a buyer's agent, helping them evaluate and decide which of several vendor MPS proposals to invest in. It is through these experiences that I have gained and continue to gain a thorough understanding of the buyer's perspective when it comes to investigating, pursuing, evaluating, purchasing, implementing, managing, and evaluating the performance of Managed Print Services.

I have also gained valuable experience as a MPS solution provider, having placed approximately $400M in total MPS contracts over the years. The combination of this experience as a MPS provider combined with my experience as a MPS program developer and buyers' agent has given me the insight to know what is good, what is bad, what works, what does not, what will succeed, what will fail, what is bulls#@t, what is credible, and what buyers should do to maximize the chances of implementing a successful MPS program. This book is my attempt at sharing real-world, first-hand, experience-based, proven insight in hopes of helping buyers navigate the convolutions of the MPS purchase. In that regard, I hope I have succeeded.

ABOUT THE AUTHOR

Tab Edwards is a Principal with The Water Group, a Business Services Consultancy, dedicated to providing organizations with ideas, support, services, and solutions that improve overall business performance.

Edwards is widely considered to be the world's foremost authority on Managed Print Services. Since 1999, Tab has developed and supported the development of corporate and channel Managed Print Services programs (including Hewlett-Packard, for instance) and he has placed nearly $400M in true MPS solutions globally, helping buyers save more than $100M on their cost of ownership.

His passion for helping organizations and their associates is reflected in his consulting, speaking, and writing. He is a frequently sought-after speaker and the author of ten books, including the Amazon.com Bestseller *MPS: Managed Print Services*, *Paper Problems*, and *Imaging & Output Strategy*.

When Edwards is not writing, he consults with business leaders across multiple industries and countries, helping them to do better business, including delivering better performance through Imaging & Output solutions such as Managed Print Services.

✖️ thewatergroup

The Water Group is dedicated to helping organizations of all kinds function more effectively through planning, execution, and associate development.

Consulting: Water Group Consultants employ The Water Method framework that we call "P.E.M.A." (Plan. Execute. Monitor. Adjust) in all our consulting and training engagements. With a variety of service offerings, our engagements are practical and application oriented.

Speaking: Whether it's a keynote address, sales seminar, motivational event, or inspirational, Tab Edwards is considered to be among the absolute best, most effective, engaging, and entertaining public speakers and trainers.

Books: Tab Edwards' ten books, including an Amazon.com Bestseller, tackle topics surrounding strategy, business process improvement, Imaging & Output effectiveness, Managed Print Services, sales, and personal growth.

Software: Sales Navigator +lus (SNaP) and other Water Group tools were developed to help organizations and professionals improve performance and results.

The Expert Lounge: As a courtesy, The Water Group provides a free service for business professionals seeking expert advice in one of the areas of our team's expertise, including strategy, business process improvement (including MPS), and sales effectiveness.

INDEX

20 Questions 128

A

A3 format 63, 80
Access Control 150
ACPP 50, 51, 52, 62, 64, 95
Action Plans 39
Adjustment activities 124
Also by Tab Edwards 4
  Art of Movement 4
  Batman, Robin, David
      Beckham,
  and the Naked King 4
  Chocolate Peppers 4
  Coffee is for Closers ONLY!
      4
  I&O: Imaging & Output
      Strategy 4
  Lessons of the Navel Or-
      ange 4
  MPS: Managed Print Ser-
      vices 4
  Paper Problems 4
Amazon.com Bestseller 157
American Paper Institute 109
Analyze the Current Bench-
      mark State 62
Art of Movement 4
assessment 42
  Benchmark 42
Assessment 42
Assessment Data Gathering
      104
  Adjustment activities 124
  Automated approach 105
  Manual approach 105

Inventory Turnover Rate
      119
  Manufacture date 121
  Pros and Cons 124
  Statistical error in each
      data-gathering 107
  Usage cycle 108
Assessment Defined 42
Assessment: Outcomes from
      the assessment effort
      89
  Building the Business Case
      90
  Collaboration 90
  Development of the full list
      of TCO factors 90
  Findings & Recommenda-
      tions Report 89
Asset Management 48, 86
Automated approach 105,
      107, 127
Automated Data Collection
  Automated approach 108
  Manual Data Collection
      118
Average Cost-Per-Page 50
Average (Formula) 123
Average Inventory 120

B

Balancing the device deploy-
      ment 64
Base Plus Click 82
Batman, Robin, David Beck-
      ham,
  and the Naked King 4

Being more "Green." 35
Benefits of MPS to the Buyer 31
Billing models 82
  Base Plus Click 82
  Level-Pay Billing 83
  Pre-Paid Pages Plus Overages 84
  Pure Cost-per-Page 83
Biosphere 35, 36
Building the Business Case 90, 91, 140
  Cost-benefit analysis 91
  Implementation timeline 91
  Potential risks 91
  Purpose of the project 91
  Results-validation 91
  Technology 91
Buyer Bad Habits 143
  False perceptions 147
  Fragmented Imaging & Output responsibility 144
  Lackluster internal communication 146
buyer enlightenment 49

C

Canon 26
cash 58, 59, 71
Clarifying Managed Print Services 29
Coffee is for Closers ONLY! 4
Collect Cost Data to Calculate the Total Cost of Ownership 46
Collect Quantitative & Qualitative Data 56
Collect User Input and Feedback 60
competing printers 54
Confidence level 114
Conservative Paper Consumers 109
Considerations 135
Consolidating devices 64
construction industry accounting 58
Continuous Process Improvement 76, 77, 142
  CPI 76
  Fleet Management and Continuous Process Improvement 77
Contract
  MPS agreement 78
  technology refresh 79
Cost-Benefit Analysis, 67
  Internal Rate of Return 67
  Net Present Value 67
  Payback Period 67
  Return on Investment 67
Cost factors 47
Cost of Goods Sold 120
Cost-per-page 49
Cost-reduction Distribution Diagram 103
Cost-reduction drivers 103
Cost-Reduction Savings Distribution 103

costs 101 58
Cost Savings Illusion 140
CPI 76
CPP 49
creation of a strategy 32
Current Benchmark State
        21, 23, 33, 62, 63, 66,
        67, 71, 89, 91
Current State Snapshot 69

D

Data Gathering Methods 124
  Pros and Cons 124
definition 31
Deployment Project Manage-
        ment 87
Design a Solution for Im-
        provement (Recom-
        mended Future State)
        63
Designing efficient user-re-
        lated workflow models
        64
Determine the "Study
        Group" 42
Develop a Current Bench-
        mark State vs. Recom-
        mended Future State
        Comparison 66
Develop a Topological Map
        60
  Topological Map 60
  topology 60
Device and output consolida-
        tion 94
Device Consolidation Dia-

gram 96
device-to-user ratios 64
Direct Costs 57
Discovery & Design 92

E

Elements of a True
Managed Print Services Solu-
        tion 37
  Assessment 42
  Fleet Design 70
  Hardware 73
  Imaging & Output Strategy
        37
  MPS Service Options 86
  Program Management 75
  Service & Support 74
  Single Invoice Billing 81
  Supplies and Consumables
        73
  Tracking and Reporting 80
Energy 35, 36
Essential Elements of MPS-
        Diagram 85
Existing Fleet Administration
        86
Expected Opportunity Loss
        142

F

Facilities 49
Financing arrangement 137
  Master Lease / Finance
        Lease 138
  One-Dollar Buyout 137
  Operating Lease 137

Findings & Recommenda-
    tions Report 89
Fleet Design 70, 71, 72
    Fleet Design Sample 72
    Solution Design 70
Fleet downsizing 77
Fleet Management 77, 93
    Fleet downsizing 77
Fleet Management and
        Continuous Process
        Improvement 77
foldin' money 59
Forrester Research 17
Fragmentation 144
Freddy Krueger 38
Freedom to focus on core
        competencies 35
From Where the MPS Cost
        Savings Emanate 94
    Device and output consoli-
        dation 94
    Device Consolidation
        Diagram 96
    Improved device reliability
        and uptimes 97
    Waste-reduction through
        more efficient pro-
        cesses 100

G

Global Warming 35, 36
Goals 37
Green 35
    Biosphere 35
    Energy 35
    Global Warming 35
    Preservation 35
Green initiatives 35

H

Hardcopy Operational Assess-
        ment 106
Hard Costs 57
    cash 59
        foldin' money 59
Hard-dollar cost savings 32
Hardware 73
Harvard-educated mechanical
        engineer 41
helpdesk 74
Helpdesk
    Printer-related helpdesk
        calls 99
Helpdesk support calls 99
Hewlett-Packard 26, 157
How Costs are applied to
        each device to deter-
        mine the device's TCO
        and Average Cost-per-
        Page 49
How Managed Print Services
        (Generally) Works 88
How MPS Works Diagram
        93
Human Resources Manager
        59

I

Imaging & Output 31
    Value Spectrum 132
        Value Spectrum Diagram
        134

Imaging & Output Office Assessment 42
  Analyze the Current Benchmark State 62
  Collect Cost Data to Calculate the
  Total Cost of Ownership 46
    Total Cost of Ownership Cost Factors 48
    Cost-per-page 49
  Collect Quantitative & Qualitative Data 56
    Costs 101 58
    Direct Costs 57
    Hard Costs 57
    Indirect Costs 57
    Primary costs 57
    Qualitative data 56
    quantitative data 56
    Secondary Costs 57
    Soft Costs 57
  Collect User Input and Feedback 60
  Design a Solution for Improvement (Recommended Future State) 63
    Balancing the device deployment 64
    Consolidating devices 64
    Designing efficient user-related workflow models 64
    walking paper 64
  Develop a Current Benchmark State vs. Recommended Future State Comparison 66
    Cost-Benefit Analysis, 67
    Device-to-user ratios 64
    Networking all networkable devices 64
    Reducing personal (non-shared) printers 64
    Reducing standalone analog fax machines 65
    Replacing devices aged 5-years or older 65
  Determine the "Study Group" 42
    representative study group 43
    usage characteristics 43
    Statistical Power 44
  Develop a Topological Map 60
    Mapping 60
    Topological Map 60
  Perform a User-Related Workflow Review 61
    Time & Action 61
  Present the Findings 67
Imaging & Output Strategy 4, 37, 157
I'm just gonna swim! 42
Implementation of the MPS 92
  Discovery & Design 92
  implementation timeline 91
  Improved device reliability and uptimes 97

Improved user productivity 34
Increased user satisfaction 33
Indirect Costs 57
Inexperience. 23
Initiatives: 39
Installation / De-Installation 87
Internal Rate of Return 67
INTRODUCTION 15
Inventory Turnover Rate 119
I&O Assessment Cycle 68
I&O: Imaging & Output Strategy 4
Irrational client expectations 24
I/T managers 49

K

Knowledge 34

L

Lackluster internal communication 146
Lack of client preparedness 24
Lack of consistent global delivery capabilities 26
    local-country partners 26
    Morocco 26
Lack of consistent global delivery capabilities. 26
lake 41
Lessons of the Navel Orange 4
Level-Pay Billing 83

Lexmark 26
local-country partners 26
lower TCO than the competitors' printers 53

M

Managed Print Services 17, 30
  Benefits of MPS to the Buyer 31
    Being more "Green." 35
    creation of a strategy 32
    Freedom to focus on core competencies 35
    Hard-dollar cost savings 32
    Improved user productivity 34
    Increased user satisfaction 33
    Knowledge 34
    mproved reliability & uptime 33
    Optimization 34
  definition 31
  Elements of a True
Managed Print Services Solution 37
    Fleet Design 70
      Solution Design 70
    Fleet Management and Continuous Process Improvement 77
    Hardware 73
    Imaging & Output Strategy 37

MPS Service Options 86
Program Management 75
  Continuous Process
    Improvement 76
  Service & Support 74
  Single Invoice Billing 81
    Base Plus Click 82
    Level-Pay Billing 83
    MPS financing 137
    Pre-Paid Pages Plus
      Overages 84
    Pure Cost-per-Page 83
  Supplies and Consum-
    ables 73
  Tracking and Reporting
    80
Imaging & Output-based
    solution 30
reasons why Managed Print
    Services implementa-
    tions fail 60
  users do not accept the
    solution 60
Manual approach 105, 106,
    107, 118, 122, 126
Manual Data Collection 118
  Inventory Turnover Rate
    119
  Manufacture date 121
Manufacture date 121
Mapping 60
Margin of error 114
Master Lease / Finance Lease
    138
McKinsey 38
mean-time between failure

    65, 97
measuring success 18
Morocco 26
Movement 7
mproved reliability & uptime
    33
MPS agreement 78
MPS Cost Savings 94
  From Where the MPS Cost
    Savings Emanate 94
MPS financing 137
MPS: Managed Print Services
    4
MPS Service Options 86
  Asset Management 86
  Deployment Project Man-
    agement 87
  Existing Fleet Administra-
    tion 86
  Installation / De-Installation
    87
  Ongoing Optimization 87
  Phone Support 86
  Software 86
  Training / Transition Man-
    agement 87
multifunctional devices/pe-
    ripherals (MFPs) 30
Multi-Year Cost Savings Illu-
    sion 140
  Expected Opportunity Loss
    142
mutual funds 58

N

Net Present Value 67

Networking all networkable devices 64

O

Objectives 39
One-Dollar Buyout 137
Ongoing Optimization 87
Operating Lease 137
Optimization 34
ost factors 47
outsourcing solution 17

P

Paper consumption 119
Paper Consumption Spectrum 108, 109, 110
Paper Problems 4
Paper Waste 97
Payback Period 18, 67
Perform a User-Related Workflow Review 61
Personal Printers 65
Philadelphia 8
Phone Support 86
Poorly-conducted office assessment 23
Pre-Paid Pages Plus Overages 84
Present the Findings 67
Preservation 35
pricing models 82
Printer A and Printer B 54
printer-related helpdesk calls 99
printers aged 5 years or older 65

Print Security 148
  Hacking and re-routing 149
  Network penetration 149
  Network sniffing 149
  Print output 149
  Securing the Imaging & Output environment 150
    Access Control 150
    Document security 150
    Monitor and track the device fleet 150
    Protect information on the network 150
    Secure the product 150
Probability 112
  Confidence level 114
  Margin of error 114
  Mean 114
  Randomly sampling 115
  Sample mean 113
Program Management 75, 92
Prosperous Paper Consumers 109
Purchasing 49
Pure Cost-per-Page 83
Purpose of the project 91

Q

Qualitative data 56
quantitative data 56, 62

R

Randomly sampling 115
Real Estate 49
reasons why Managed Print

Services implementa-
tions fail 60
users do not accept the solution 60
Recommended Future State
63
Recommended Future State
design 63, 71
Reducing personal (non-
shared) printers 64
Reducing standalone analog
fax machines 65
Replacing devices aged
5-years or older 65
representative study group 43
Responsibility Matrix 145
results-validation 91
Results Validation 20, 21, 22,
25, 32, 103
Results Validation process 20
Return on Investment 18, 67,
139
Ricoh 26

S

Sampling Error Diagram 116
Securing the Imaging & Out-
put environment 150
Security 126, 148
Hacking and re-routing 149
Network penetration 149
Network sniffing 149
Print Security 148
Securing the Imaging &
Output environment
150

Access Control 150
Document security 150
Monitor and track the
device fleet 150
Protect information on the
network 150
Secure the product 150
Service & Support 74
availability 74
fleet uptimes 74
helpdesk 74
preventative maintenance
74
SLA 75
timely service 74
Single Invoice Billing 81
SLA 75
Soft Costs 57
construction industry ac-
counting 58
mutual funds 58
Software 86
Solution Design 70
Sourcing 49
Statistical error in each data-
gathering 107
Statistical Power 44
strategy 32
Action Plans 39
Task List 39
Freddy Krueger 38
Goals 37
Harvard-educated mechani-
cal engineer 41
I'm just gonna swim! 42
Initiatives: 39

lake 41
Objectives 39
Strategy 4
Supplies and Consumables 73

T

Tab Edwards 8
Task List 39
technology refresh 79
Time & Action 61
Time & Action study 61
toner fulfillment 102
Toner fulfillment 102
Topological Map 60
topology 60
Total Cost of Ownership 19, 46, 50
  Average Cost-Per-Page 50
  Cost-per-page 49
  lower TCO than the competitors' printers 53
    Printer A and Printer B 54
Total Cost of Ownership Cost Factors 48, 51, 126, 127
Total Cost of Ownership Definition 46
  buyer enlightenment 49
Tracking and Reporting 80
Tracking software 118
Training / Transition Management 87
Transition Management 87
two competing printers 54

U

Unfavorable contracts 27
Usage characteristics 43
Usage cycle 108
Users do not accept the solution 60
User satisfaction with MPS 34

V

Value Spectrum 11, 132, 133, 148
Value Spectrum Diagram 134

W

Walking paper 64
Waste-reduction through more efficient processes 100
Water Group 8, 103, 109, 157
Weight-loss guru 18
What is "Managed Print Services"? 30
Where Paper Goes - Diagram 98
Why MPS Engagements Fail 22
  Inexperience. 23
  Irrational client expectations 24
  Lack of client preparedness 24
  Lack of consistent global

delivery capabilities 26
Poorly-conducted office as-
    sessment 23
Unfavorable contracts 27

X

Xerox 26

www.ingramcontent.com/pod-product-compliance
Lightning Source LLC
Chambersburg PA
CBHW050506210326
41521CB00011B/2352